Achieving Total Value in the Built Environment

Achieving Total Value in the Built Environment

Greg Watts

Jason Challender

WILEY

This edition first published 2026

Registered Offices
John Wiley & Sons, Inc., 111 River Street, Hoboken, NJ 07030, USA
John Wiley & Sons Ltd, New Era House, 8 Oldlands Way, Bognor Regis, West Sussex, PO22 9NQ, UK

For details of our global editorial offices, customer services, and more information about Wiley products visit us at www.wiley.com.

The manufacturer's authorized representative according to the EU General Product Safety Regulation is Wiley-VCH GmbH, Boschstr. 12, 69469 Weinheim, Germany, e-mail: Product_Safety@wiley.com.

Wiley also publishes its books in a variety of electronic formats and by print-on-demand. Some content that appears in standard print versions of this book may not be available in other formats.

Library of Congress Cataloging-in-Publication Data has been applied for:

Print ISBN: 9781394258550
ePDF ISBN: 9781394258574
ePUB ISBN: 9781394258567
oBook ISBN: 9781394258581

Cover Design: Wiley
Cover Image: © bubaone/Getty Images

Contents

About the Authors

Dr Jason Challender PhD, HonD Eng, MSc, FRICS, FHEA, FAPM has acquired over 36 years of 'client side' experience in the UK construction industry and has successfully procured numerous major construction programmes during this time. He is a Senior Lecturer at Liverpool John Moores University and former Visiting Professor at the University of Greater Manchester. In his current role at LJMU, he has designed two housing modules, which he will be teaching as part of a new MSc programme at the University. He is also a former Director of Estates and Facilities at the University of Salford, where he was a member of Senior Leadership Team and responsible for overseeing a large department of approximately 350 estates and construction-related staff. He is also a construction researcher, with 12 published academic journal and conference papers, as well as 10 books (mostly with Wiley), all of which have been dedicated to his studies around estates and construction management. Furthermore, he has previously served as a book reviewer for Wiley. Over the years, he has attended many national construction and institutional conferences as a guest speaker. He is also a Fellow and former Board Director of the Royal Institution of Chartered Surveyors and the North-West Construction Hub.

Dr Greg Watts EngD, MBA, PGCert, PCAP, BSc (Hons), SFHEA, MRICS has worked across the construction industry in a range of roles and for various contractors for over 10 years, before moving into academia nearly a decade ago. He is a Reader in Construction Management at Manchester Metropolitan University and has taught on numerous undergraduate and postgraduate courses. In his research, he has published extensively in articles, conference papers, journal papers and textbooks. His teaching and research interests include social value, ethics, continuing professional development, business responsibility, risk management, mega-projects, digital twinning, procurement, and mental health in the built environment.

About the Authors

Part I

Value Strategy

1

Introduction

Value is ethics. It is behaving ethically to everyone. This could be a colleague, a neighbour, a consultant, a client or a contractor; it could be towards your employer and your employees. Extending to all suppliers and supply chain partners, as well as all those directly and indirectly impacted by your own decisions. Being aware of value, what it means, the types that exist, how it is created and how it can be damaged, is key to ensuring value can be maximised. These are all topics that feature in this book. This book is essentially about value in the Built Environment (the title may have given that away already). It reflects on how we (as an industry) have arrived at the position we are in – whereby we finally have a fourth criterion to challenge the historic triumvirate of cost, time and quality. This fourth criterion often takes on different personas and hides under different guises. It goes by different names. It has been called corporate social value, sustainability, social value, health and well-being, ethics, amongst others. For the purpose of this textbook, this is called 'Total Value'. Society generally is now expecting more from all public and private sector organisations, these additional diverse demands are driving organisations to consider how best they can evidence the value they deliver. Sometimes clients are prescriptive over the value type they desire, other times they leave the finer details of value delivery down to the organisations they are procuring. Either option ultimately results in the same outcome – the need for an organisation to create, deliver, and report on value.

This book also looks forward. It considers where this new value driven focus will lead us, as an industry, and how we all, as individuals and organisations, are best placed to respond to the challenges the future will present. Whatever terminology is used in future as concepts continue to evolve, a focus on the 'Total Value' organisations bring will be increasingly judged. It will be no longer acceptable for organisations to simply donate to good causes, publish standard and non-specific policy statements and view value creation as an 'added extra' to what they do, all the while maintaining a business as usual approach that causes more harm to society than good. Organisations will be judged on their complete contribution to society and the environment – on the 'Total Value' they bring.

In Chapter 2, the concept of value is introduced. It is complex and often misunderstood due to its subjective nature. Traditionally associated with project objectives like cost, time and quality, Value has evolved to reflect a broader set of interests including those of internal and external stakeholders. Because different stakeholders, such as clients, users, designers, contractors and subcontractors, perceive value differently, it can be interpreted in a variety of ways. Whilst environmental value arguably took precedence at the end of the last century, social value has recently emerged as the main focus. However, this is still evolving and a new concept is emerging, one of 'Total Value'. This includes (and encompasses): Well-being value, Financial value, Corporate value, Employee value, Social value, Cultural value, Environmental value, Sustainable value, Reputational value, Stakeholder value and Ethical value (amongst many others). Overall, as evolution continues, value today (and more importantly, tomorrow) encompasses much more than financial metrics, reflecting broader human, environmental and ethical considerations.

Chapter 3 discusses value in the built environment. The built environment encompasses all aspects of managed space and infrastructure development, from design and construction to facility management and property investment. Unlike manufacturing, the construction industry operates through temporary teams (designers, consultants, workers, etc.) assembled for specific projects. This fragmented nature can hinder relationship-building and trust. Globally, the construction industry is significant in terms of employment and economic contribution. However, despite its scale, the industry is often berated for its struggles to modernise, which are often hampered by a focus on cost, time and quality (the 'iron triangle'), and marred by unethical practices and a poor reputation. This chapter discusses the context of the construction industry, some of its inherent problems and measures to consider value as the mechanism to drive future improvements. The legislation, some recent reforms and other governance mechanisms to drive value and act as a catalyst for improvements are then outlined.

Chapter 4 is still part of the 'value strategy' section of this textbook. The chapter introduces frameworks as a concept, how the term is used to describe multiple similar structures and also how the term is used more widely to describe a set of governance and best practice rules. A series of frameworks are then presented and discussed, with a focus on the different types of value creation each of the frameworks aligns with. The benefits and drawbacks of aligning with the different frameworks are also discussed, as are the limitations of the frameworks themselves. However, ultimately, selecting the right framework can help bring an understanding of value-based decision-making to all interested stakeholders.

In Chapter 5, the relationship between total value and a range of theoretical lenses is explored. By utilising a theoretical lens, the achievement of total value can be better understood in the context of stakeholders and how it is perceived, why it is considered legitimate (and how this legitimacy can be abused), as well as the positive contribution it can make to broader society as well as the wider economy. A new theoretical concept of Total Value theory is proposed to help better understand the modern day interaction and expectations between organisations and society. Total Value theory describes when there is a societal or market based shift from simply viewing value as an 'addition' to the 'core' function of a contract and where the entire organisation is reviewed for its overall value creation, from all its actions undertaken.

Chapter 6 discusses a few of the technologies that are currently impacting the Built Environment and will impact the built environment in future. Despite the varied use and application of the technologies, they all have many similarities. Most notably they all have the potential for industry changing Total Value creation. If developed and utilised correctly, each technology described can create efficient, effective and revolutionary value on a local and global scale. However, all the technologies come with both immediate and longer-term risks. Risks that may undermine any value achieved in other areas. The technologies are also often interrelated which brings both advantages, such as how progress in one area can support progress in another, but also disadvantages in that a failure in one area of technology development could undermine the value created by all other technology developments.

Chapter 7 is the start of part II (Value Action) and is focused on the measurement of value. It seeks to understand and capture the importance individuals, groups or organisations place on well-being-related changes resulting from specific actions. This then enables businesses to assess the impact of their decisions, demonstrating value for money and ensuring that outcomes are meaningful and effective. Due to its maturity and contemporary focus, social value is discussed on how it focuses on long-term, life-improving outcomes, which are often referred to as 'soft outcomes' due to their intangible nature. Various methods are used to measure social value of both a qualitative and quantitative nature. These tools help justify spending and attract funding but many fail to capture the essence of social value and instead provide a set of proxy values that actually serve to undermine social value creation and measurement.

Chapter 8 outlines how communicating value is as essential as creating it, especially in today's competitive, value-driven environment. Success often hinges not on the scale of impact alone, but on an organisation's ability to clearly and authentically express the benefits it delivers. However, missteps, such as unclear messaging, poor platform choices or inaccurate measurements, can undermine efforts. The most damaging of these is greenwashing, which not only erodes trust but risks discrediting organisations genuinely committed to delivering environmental or social value. Ultimately, this chapter reinforces that meaningful value exists not only in high-profile projects but also in consistent, everyday practices, so long as they are communicated truthfully. By applying the principles and strategies outlined here, organisations can ensure that their value is not only created but also recognised, understood and celebrated across all relevant stakeholder groups.

Chapter 9 introduces part III: value reflection and explores the key challenges affecting project success in the global construction industry, such as its fragmented, temporary nature and the complexities introduced by differing international regulations, cultures and behaviours. The chapter then examines the difficulties in achieving and measuring success, particularly due to differing stakeholder perspectives. Alternative models for evaluating project success are presented, focusing on project efficiency, customer impact, business outcomes and future readiness. Key performance indicators (KPIs) and their use in performance assessment are introduced, followed by a discussion of Critical Success Factors (CSFs) grouped into comfort, competence, commitment and communication. The chapter concludes by addressing the dilemmas of performance measurement on construction sites, including considerations of quality assurance, site supervision and safety.

In Chapter 10, the history of value management is discussed and how it encompasses international perspectives. The facilitation of value management is introduced and examined, as are the methodologies for measuring output. The CSFs for value management are discussed, which include having a multidisciplinary VM team, competent VM facilitator, commitment of all stakeholders and effective communication amongst participants. The benefits of value management processes are then examined from multiple perspectives. A model for identifying and tracking value is then described which encompasses value aspects from industrial, organisational, societal, professional, project and individual values perspectives. Finally, value frameworks for capturing project success are discussed.

Chapters 11 and 12 continue the theme of value reflection with personal reflections of projects and industry. The chapter defines reflective and reflective practice, and include Gibbs Reflective Cycle, Kolbs model of Reflective Practice and The Driscoll Model of Reflection. These chapters then focus on the importance and benefits of reflective practice for projects and articulate suggestions for reflective practitioners.

Chapter 13 outlines the history of value in the built environment. The idea of value, that first emerged under the guise of 'business responsibility' in the 1950s, has been on an evolutionary journey since this time. Each of this chapter's sections talks through the evolution of the value concept covering a different decade from the 1950s to the 2020s. Each new decade brings to the concept new ideas and expectations. It is no longer enough to continue business operations as normal, and treat value based behaviour as an additional organisational operation. Value needs to be intertwined with all operations. A failure to do so will lead to society viewing an organisation as outdated and obsolete, with no place in modern society.

Finally, Chapter 14 outlines several scenarios of potential value creation opportunities. Questions are posed which serve to challenge assumptions and provoke ideas on how value can be achieved and delivered by multiple stakeholders. The scenarios and accompanying questions serve as practitioner and student examples, and the answers provided help illustrate the opportunities (and potential drawbacks and conflicts) a value driven approach can create.

The main aim of the textbook is to introduce the concept of Total Value. To do so, a robust background of value is provided, and an introduction to the different forms of value that can be created and delivered. Essentially, Total Value theory describes when there is a societal or market based shift from simply viewing value as an 'addition' to the 'core' function of a contract and where the entire organisation is reviewed for its overall value creation, from all its actions undertaken. It is no longer acceptable for organisations to have negative consequences and implications for society, no matter how many 'additional' initiatives it undertakes to create value for society. If the organisation at its 'core' provides services that of detrimental to society, they will have to evolve their purpose quickly or risk falling into history as an organisation that could not provide society with the value it expected. This textbook will aid students, academics and industry practitioners with the skills and knowledge needed to truly understand Total Value. With a greater understanding comes the ability to transform society and the planet with a value focused approach that is meaningful, impactful and transformative.

2

What Is Value?

Nowadays people know the price of everything and the value of nothing.
Oscar Wilde's quote about value from The Picture of Dorian Gray

2.1 Introduction to Chapter

The above quotation by the notorious Irish poet is symbolic of the natural tendency and propensity for society to be fixated on price and cost rather than what provides real value. The reasons for this can be varied, but the main dilemma may be that people are confused by what value represents and simply regard it as the cost of something. Value is clearly more than just financial and a differentiation between price and value is therefore clearly required.

This chapter of the book responds to the above dilemma and will begin by offering context of what constitutes value from different perspectives. It will define value and explain the challenges that this can present from providing a universal definition, given that value means different things to different people. The chapter will then look at a brief history of value from the era of the Greek philosophers to give a perspective on where it emanated from and how its concept has been developed over the centuries. The different types of values, as illustrated in Figure 2.1, will then be identified and explained, and the main differences between them will be articulated.

Within some of the categories of value, there will be practical examples provided to assist readers in understanding how value can be derived in each case. Furthermore, there will be a narrative provided which is aimed at distinguishing and differentiating each category of value from another.

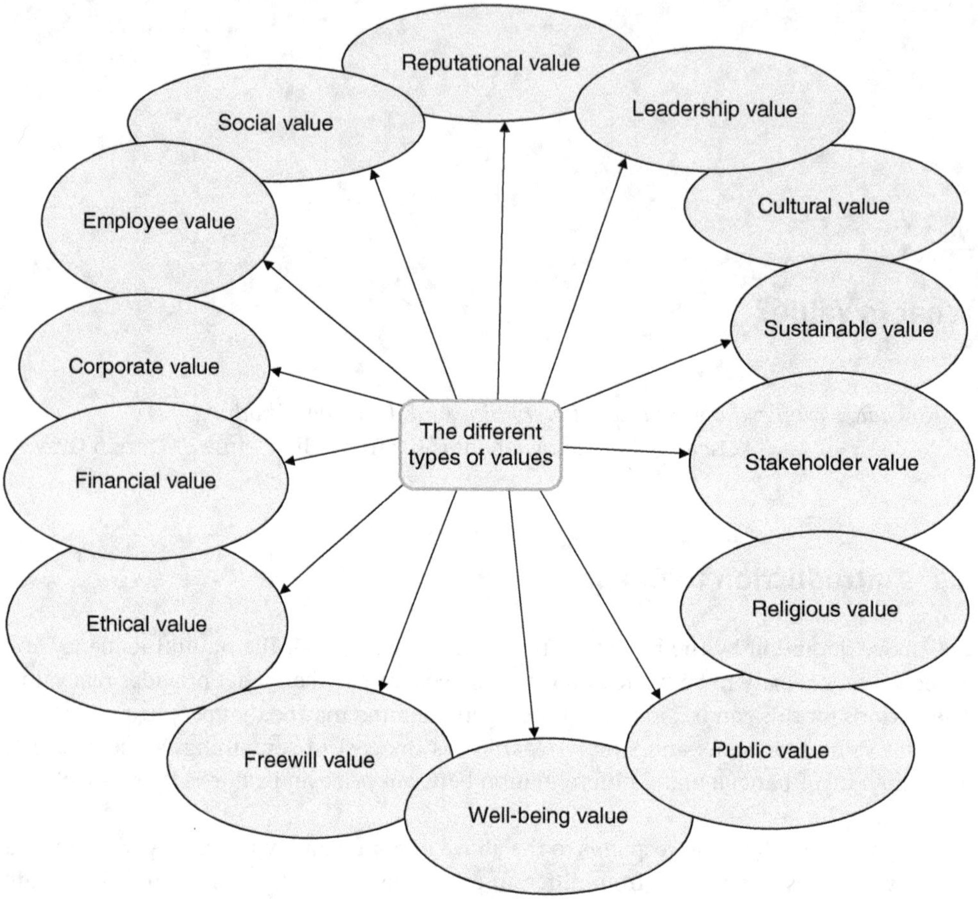

Figure 2.1 The main categories and types of value.

2.2 The Notion of What Constitutes Value

The notion of value in the construction industry is a subject that has received much debate for many years since its initial conception. According to Khalife and Hamzeh (2020), the concept of value is one which has been misinterpreted and commonly overused in the past. Traditionally value has been linked to the objectives and needs of project owners (commonly referred to as project sponsors) and clients. Notwithstanding this assertion, project value has become increasingly used across a broader spectrum to represent other internal and external stakeholders' visions, requirements and needs on projects. These can include the economic, environmental and social requirements of stakeholders and the public.

Value has been described as a notion that is not static and that can change with time owing to its dynamic nature (Thyssen et al., 2010). This is mostly attributable to changes brought about by the interaction of the different project attributes which will naturally develop as projects progress through time. Furthermore Kliniotou (2004) advocated that value is commonly misunderstood and can vary between different individuals. People will normally focus on value from the perspective of seeing it in the environment they operate within and in what they do. Accordingly, the different internal and external stakeholders, including clients, end users, design consultants, contractors and subcontractors, will see value from many different viewpoints and spectrums. According to Kliniotou (2004), it is important to determine and assess value away from the traditional 'iron triangle' related to cost, time and quality outcomes. Notwithstanding this premise, project performance should also be considered from other angles to encapsulate value performance. This is especially relevant at a time when the global construction industry is going through a systematic transgression into more modern practices and ways of thinking. In this pursuit, Fowler (1990, as cited in Khalife and Hamzeh, 2020) created a formula for calculating value. He described and expressed value as the ratio of worth over cost or alternatively function and quality over cost.

There may be a subjective nature to value in this regard, as value may be strongly influenced by the initial impressions of end users in use over both the initial capital costs and the cost in use associated with operational cost, as part of life cycle analysis. To add further debate as to the concept of what constitutes value, some believe that it represents the balance and relationship between both positive and negative consequences.

2.3 A Brief History of Value

The first documented transcript of what value means came from the Athenian Academy around the 4th Century BC. At this time, Aristotle found that value was related to need from an exchange of goods and services perspective. He decreed that, without need, there would be no exchange potential and therein no value. In this sense something could have a positive value if it provides a contribution to being able to satisfy certain needs and requirements. In addition, he advocated that it could hold value if it decreases deficiencies or drives or improves human survival. An alternative theory of what value creation comprises emerged in the late 19th century. This traditional perspective on value found that it is founded on the cost of labour and resources that went into the creation of something.

The first reference recorded about value management (VM) dates back to as early as 1731, when the academic of his time, namely Daniel Bernoulli, prepared a research paper for the St. Petersburg Imperial Academy of Sciences. The paper presented a case that the value of something should not be purely based on its cost or price, but alternatively on the benefits that it brings. The concept of value management will be discussed more intensely later in this chapter.

2.4 Explanation of the Different Types of Value

In today's society, there are increasing references to 'value' and 'value creation' from many different angles and perspectives. It can sometimes be challenging to fully comprehend what value means in this context, especially as it can mean different things to different people. Accordingly, this chapter will attempt to differentiate and distinguish the main categories of value which represent wide ranging aspects from a social, economic and financial, ethical, reputational and environmental standpoint.

2.5 Financial Value

The financial value of something is the price which a willing purchaser would be prepared to pay to a willing seller for something in an unregulated market.

In terms of accounting, financial value represents the monetary worth of something which could be an asset e.g. property, goods bought or sold, business entity and services undertaken. The concept of value from an economic perspective represents the cumulative sum of all the rights and benefits resulting from ownership. In some cases, financial value has been seen as a positive entity, but there are cases where it could creep into negative territory. This is particularly the case where it could include any liability or obligation acquired. An example could be where a householder has 'negative equity' where the financial value of their home is valued and worth less than the mortgage that is held over it.

An alternative definition of financial value is presented below:

> *The usefulness or significance of records containing financial information that is necessary to conduct current or future business or that serves as evidence of financial transactions.*

In addition, according to Maher (1992), financial value can be related to some documents that record the monetary value of property held or goods and services received in exchange for funds. Such records have financial value if they are necessary to account for property, receipts or expenditure later relied upon by audits and governance regulations. This is especially the case for public bodies who have duties to keep proper record keeping for examination and evaluation at any time in the future.

It is common in the corporate world, when considering financial value, to use a *Price to Earnings Ratio Valuation* method to determine the stock price of an organisation against the potential profit that any investors can expect to gain from it. This is frequently calculated by averaging both the earnings and share price over a given year. Determining financial value is important as it can assist individuals and companies in budgeting and clearly giving a comprehensible position on both income and expenses.

Sometimes, financial value can be linked and associated with net worth. For individuals net worth represents the most significant number from a financial perspective. Net worth is essentially the difference between what people own and what they owe. Accordingly, it can be calculated by deducting the total of all personal liabilities from the value of one's assets, giving a perspective of the financial health of an individual. In conducting valuation within a business context, financial value would be calculated in a similar way. It would include considering the value of assets the business owns, such as property, plant and equipment. However, the value of a business is not solely related to net assets and a business's balance sheet, containing information on liabilities and earnings, would only represent the starting point in the valuation process. Other factors will need to be considered, and these will be discussed later in this chapter.

When considering financial value could be related to a non-physical asset that generates its value from an ownership claim or contractual right. Its value can be heavily dependent on market forces associated with supply and demand. In determining the value of an asset, financial investment or product, one could peruse actual transactions in the marketplace for similar assets. In a domestic context, this could involve checking the sale process of similar houses in the local area. In a business context, this could entail undertaking an estimation of the expected asset earnings and determining the cost of asset replacement. In addition, company valuation normally involves an accountant working out the total revenue for the business and then deducting all operating expenses before multiplying this by an industry norm. An example is included below.

Example of a multiplier effect on financial value
A car sales business turns over £10m post-tax revenue in its latest accounts All operating pay and non-pay expenses amount to £5m The company has made a post-tax profit of £5m. If the multiplier for the value of businesses in the car sales market is, say, 11, then the business could be worth £5m × 11 = £55m.

This is a very crude way to value a business, as it could be very dependent on other factors such as future business, predicted market volatility, predicted sales and revenue, depreciation costs of its assets and potential future capital replacement liabilities. The three main methods of company valuation, following established common valuation metrics, are the discounted cash flow, precedent transactions approach and public company comparison. An explanation of each of these methods goes beyond the scope of this book.

It should be stressed that financial value is seldom a science and cannot always be calculated taking account of what has previously been referred to. As auctioneers will decree, the value of something is the price someone wishes to pay for it. Accordingly in a good market where an item is in demand by a particular buyer, then clearly the price they are prepared to pay could vastly increase.

2.6 Corporate Value

The term corporate value refers to the core principles that guide the behaviours, culture and decision-making within an organisation. Such values are widely considered to form the beliefs and motivation that can define a business's direction, purpose and identity. They should be formulated around a guiding philosophy related to improving company service levels to their customers, employees and the public at large. Sometimes the terminology corporate value can be misunderstood for the financial value of an organisation. In fact, the notion of corporate values extends far beyond financial performance. Moreover, it encapsulates moral standards and principles which the company serves to uphold which can serve to provide the right pathway for a company to proceed along. These are important factors as they demonstrate a commitment to all other things non-financial. Furthermore, they help shape the work habits and behaviours of staff and serve as promises of how a

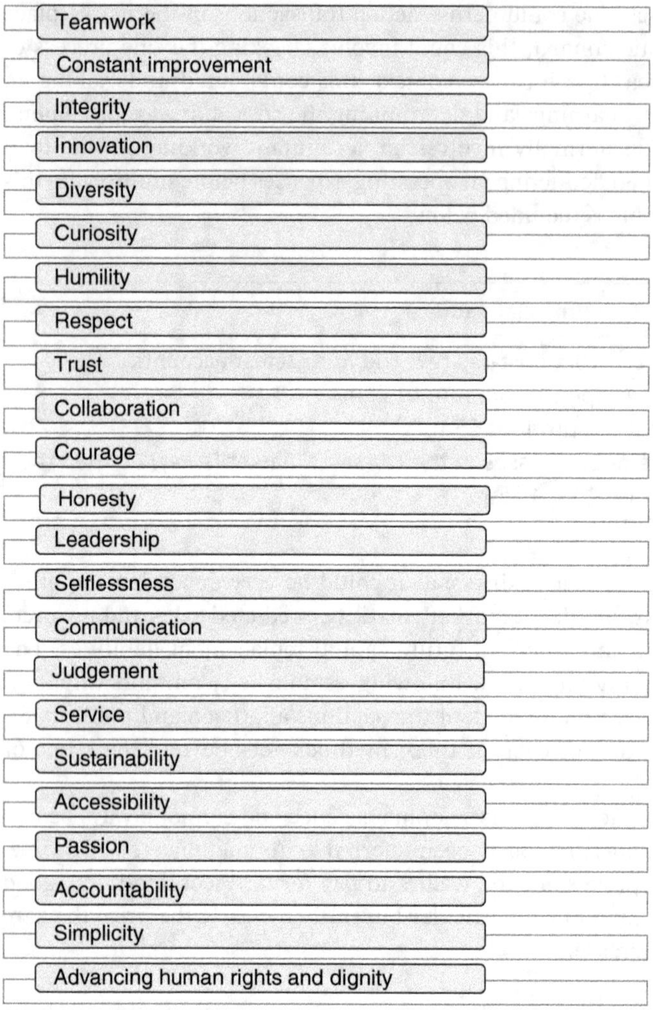

Figure 2.2 The main factors that shape corporate value.

company will treat its employees, customers, supply chain and local community. Corporate values have been described in the past as forming the DNA of an organisation. Whilst the list of factors which shape corporate value is endless, Figure 2.2 captures some of the main driving forces.

2.7 Employee Value

An employee value proposition (EVP) has been described as the statement of values, support, company culture, recognition and rewards that employers give to their employees. This is designed to provide employees to carry out their work in the most proficient way and thus go on to reach their maximum potential and gain successful outcomes. EVP is regarded as important as it can assist in differentiating an organisation in a competitive marketplace from other competitors and therefore attract the most talented staff. Employees will be drawn to companies that possess a strong EVP as it should articulate why potential candidates should join the business and how employer values, mission statement and culture are aligned to meeting employees' career aspirations. Employee value propositions also assist employers in making the right decisions by constantly asking themselves whether decision-making is echoing the values of the organisation. This should assure them that the right decisions are being made. If there is a non-alignment of decision-making with company values this is regarded as a major risk and individuals should reconsider their responses.

Employees are much more likely to remain with an organisation if there is a feeling of being valued. This notion of 'self-value' can provide benefits for employers and employees in providing a stable environment, which employees feel comfortable within and therefore perform better than they would if they feel they are undervalued or under threat. For an employer, maintaining a stable, motivated workforce has the advantage of lower turnover rates and reduces the cost to the business in terms of the recruitment process for new staff. Retention of existing staff for them also ensures that experience is not lost to their competitors and the best employees can progress upwards within the business through promotion and other career prospects. The five main pillars of employee value proposition are widely regarded as career progression and development, reward and recognition, work-life balance, purpose and mission and the culture of a company.

According to Challender (2023), those companies that have 'gone the extra mile' for their employees, through reward and recognition, commitments to staff development and measures to improve job satisfaction have reaped many associated benefits. These have included raising their ranking on schemes such as *Best Companies* in the UK. The Best Companies accreditation scheme in the UK represents a national benchmark which shows that employers are taking workplace engagement seriously. This can send a clear message out that these organisations have reached a level of significant achievement in terms of employee satisfaction. The scheme is a model based on ethical leadership which is in turn predicated on employee interaction and an unequivocal and unconditional commitment to the workforce.

There are main different ways and means to make employees feel valued in the workplace. Some of these are illustrated in Figure 2.3.

In terms of financial value there have been attempts to calculate the worth of an employee to a business. Although very crude this has sometimes been worked out by dividing the firm's net revenue by the total number of employees. However, this would not provide worth for a given employee but moreover only serve to give an average number for the worth of all employees within a business.

Wherever possible businesses should request open and honest feedback from their employees. Some of the questions might include the following:

- What values does the organisation embody?
- What values should the organisation hold?
- What values do employees have and are these aligned to the organisation?

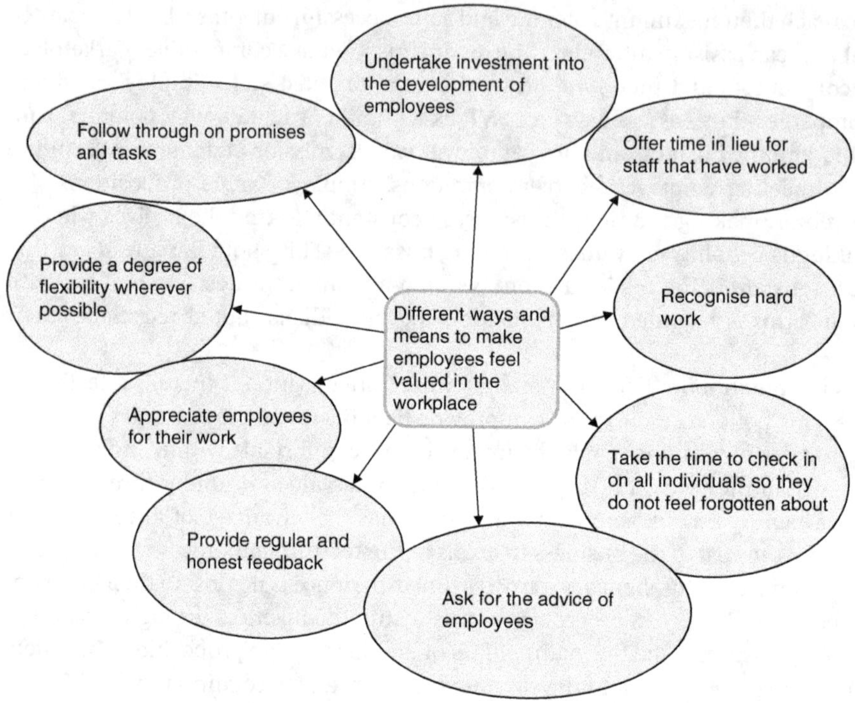

Figure 2.3 Different ways and means to make employees feel valued in the workplace.

2.8 Social Value

The UK definition of social value relates to the changed experiences that people face and the outcomes and impact of these on their lives and contributes to their overall well-being. In this sense some changes are more significant for them than others and social value is concerned with quantifying the importance and significance that change has on people's lives.

According to the British Standard Institute BS8950:2020, social value is generated from the short- and long-term collective well-being of people. Accordingly, the standard advocates that all organisations should adapt and convert some of their resources for benefitting well-being in this regard. There is an admission that social value can be created or destroyed by the actions of organisations. Accordingly, there can be outcomes which can be deemed as both positive and negative in such cases, BS8950 states that social value is the purpose and result of sustainability. In this sense, sustainability is concerned with the achievement of an optimum level of well-being (social value) in society. According to Auerswald (2009) social value is something created by social entrepreneurs that goes over and above what businesses and traditional entrepreneurs deliver. He argued that whilst conventional entrepreneurs are concerned with creating financial value, social entrepreneurs are focused on creating social value. Without such intervention, the position and state of societies and communities that we live in would remain the same with challenges not addressed. In addition, in the UK, *British Values* have been introduced which are essentially related to rules of law, democracy, respect and tolerance and individual liberty.

In the UK social value emerged from the Public Services (Social Value) Act of 2012. This was related to the way that finite resources are used and allocated. It entails that organisations should look beyond the financial value of each individual contract they undertake but alternatively focus on the collective community benefit. Social value begs the question of whether money expended on service delivery including construction could be utilised to also generate wider benefits for communities. The Act lays down responsibilities to consider those benefits as part of the decision-making processes and applies to public sector bodies for contracts above £111,676 and for other bodies above £172,514.

2.8.1 Building Social Value into the Commissioning and Procurement Processes

The ways and means to create social value can revolve around procurement strategies and giving consideration to helping access for small businesses, under-represented groups and disabled persons in the workplace. It is also concerned with environmental impact. In the case of central government funded procurement there is a requirement for social value to be explicitly evaluated rather than just simply considered under the Public Services (Social Value) Act of 2012. Accordingly, there is a requirement for some contracting and commissioning organisations to construct social value into the specification of their service agreements. In addition, the aims and visions of the organisations should be clearly set down. This should act as a golden thread across the whole life cycle of a development or service agreement.

There is normally a staged approach adopted in implementing social value initiatives on construction projects. An example of this is contained in Figure 2.4.

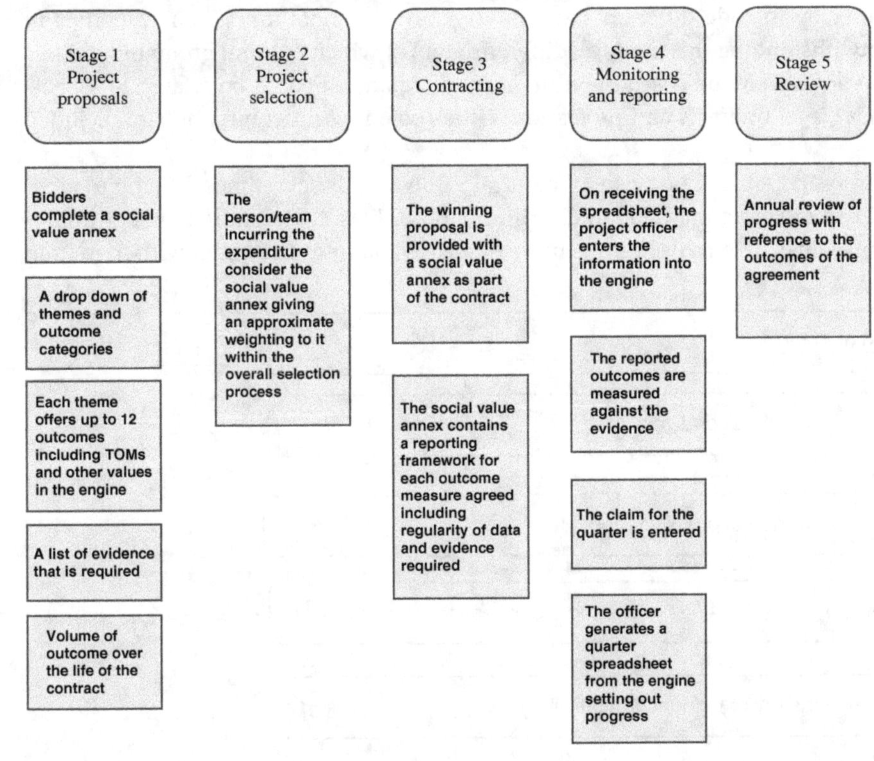

Figure 2.4 Staged approach adopted in implementing social value initiatives on construction projects.

2.9 Cultural Value

Cultural values represent the core ideals and principles upon which our communities are built on. In this sense, they represent the beliefs of a particular culture on what is right and good and are sometimes referred to as *cultural value preferences*. They are influenced by those cultures that we would normally associate with ourselves. Such values are not necessarily positive or negative but different from one another. There are many benefits from understanding different cultures. Having an appreciation of which cultures people associate themselves with should allow society to better interact with individuals around them. This knowledge and awareness can encourage respect and empathy for others and assists in celebrating differences and similarities amongst different cultures. The world that we live in is becoming more multicultural. This has been exacerbated by the increase in international travel and the use of the internet, especially around video conferencing. Accordingly, there should be no barriers for different cultures to be understood. For this reason, having a good awareness of different cultures and having the ability to associate with them, can improve communications and interactions from an international perspective. This reinforces the point that cultural values can affect how individuals and groups want to be treated and how we treat others on a global scale. Within a particular cultural group there can be differences, so it is important not to stereotype them in any way or make assumptions about them. Within a particular cultural faith such as Christianity there will be differences of how group members believe and act according to certain aspects of their religion. This may include different interpretations of the bible from their teachings in the past. Notwithstanding this premise, differentiators of which cultural group individuals form part do not have to be based on ethnicity or religion but could be related to aspects such as gender, age, interests and profession. Other potential differentiators are included in Figure 2.5.

It is not always possible to be aware of cultural values as it is sometimes difficult to identify them. This can sometimes lead to discrimination or bias. By learning to better understand cultural values it helps us become more inclusive and integrate better with those that

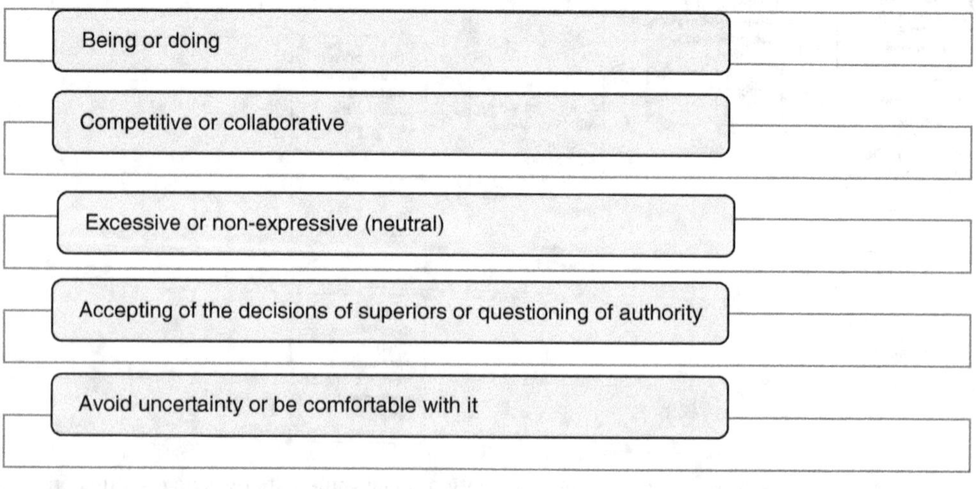

Figure 2.5 Potential differentiators of social group.

share a different cultural position to ourselves. However, it is not always the case that we are aware of our own cultural values as we take them for granted in most cases. Therefore, becoming acutely aware of our own cultural values can improve the way we relate and the relationships that we build with others.

2.10 Reputational Value

Reputational value is frequently referred to as 'brand value' and represents the perceived value of an organisation or brand. Companies can trade on their brands and use their name and reputation to increase their share of competitive markets with the intention of increasing sales, profits and dividends for their shareholders. Such well-established household names such as Coca-Cola and Apple are well known on a global scale and synonymous for the quality of their products. Reputation is crucial for businesses such as these as it has a major influence on trust and loyalty predominantly for customers. Reputation is also an important factor for developing partnerships and building good employee relationships. If a company has a good reputation, it can attract other businesses to collaborate and partner with it. This may be part of its supply chain or simply companies who wish to partner in new ventures possibly involving research and development (R and D). Conversely, those organisations that have a poor track record and have suffered reputational damage in the past may struggle to bring on board new investors or partners. They may also experience difficulties in recruiting the right staff who are more likely to join other organisations where their reputations have not faltered. Reputation is also important for creating the right operating environment. A successful company that holds an unblemished reputation should find that there is an instilled motivation to maintain high quality standards to protect the brand. This can have benefits associated with creating a more open, productive, inclusive and enjoyable working environment.

Reputation can serve as an intangible asset for an organisation and thus have an influence on the perceptions and decision-making of stakeholders. One example could be that senior managers may resist the temptation of entering particularly risky or unethical business ventures which could easily backfire and cause reputational damage. Another could be that there may be a greater propensity for banks and investors to become associated with those companies which have the highest brand value.

Notwithstanding the above premise, reputation is not something that can be guaranteed across the lifespan of an individual, organisation or product. This is explained by Challender et al. (2019) through the dependency that reputation has on the fragility of trust. There have been widespread cases in the past where individuals or organisations have been 'dropped' by sponsors for misdemeanours that have surfaced which has tarnished their reputations. In such cases sponsors will be wanting to avoid their own reputational damage in being associated with these offending individuals or organisations. In some cases, these investor sponsors have opted for damage limitation by severing all relationships with them. Examples include sports gear sponsors for famous sports personalities who have withdrawn when there have been criminal investigations levelled against them in the press.

It is the reputation of a person or company that makes them special and distinctive in the marketplace. This can be the deciding factor on whether customers will be loyal and buy a particular product or not. There have been many ways and means for assessing the reputation of an organisation. This has included stakeholder surveys including clients, customers, employees and investors. It has also included opinion polls and media analysis and discussion with focus groups and industry executives.

A good example of gaining a good reputation is by following the adage '*saying what you mean and doing what you say*'. Along these lines it involves setting out one's intentions and being true to your word, without the temptations to overcommit oneself and remaining on top of the business in hand. These are traits which are highly respected and sought after in all professions. Those companies that 'promise a lot and deliver very little' are the ones who have been found short in this area and where customer loyalty, trust, goodwill have suffered. One example of a well-known and previously respected company, namely Hoover, where its reputation in the 1990s 'nose-dived' nearly caused its demise after nearly 84 years in business is contained below.

One of the worst sales promotions in history: reputational value of Hoover severely damaged

In 1992, the vacuum cleaner manufacturer, Hoover, launched through its UK branch what could only be described as an incredibly generous promotion. It offered customers the prospect of two return flights to the USA if they bought a Hoover product over £100. Up until this time Hoover was an established household name, the market leader in vacuum cleaners and other domestic items of equipment.

The promotion was intended to represent an innovative way of bolstering dwindling sales by shrugging off an increased number of competitors and thereby 'turning the tide' on its performance during a recession. Hoover partnered with a company called JSI Travel in this initiative which was regarded as a win-win scenario for both businesses. However, Hoover was acutely aware that if everyone who purchased an eligible product applied this would carry significant risk for them. For this reason, they orchestrated a process for redemption of the offer that was designed to be time consuming and would hopefully serve to deter most customers from applying. They had also assumed that customers would buy products well in excess of the £100 qualifying threshold which turned out not to be the case.

As sales and associated application for free flights went 'through the roof' Hoover realised that they were unable to honour their promotion, with the cost of flights outweighing the sales from vacuum cleaners by several times. Accordingly, it began to implement a strategy to entice their customers out of the promotion. They attempted this by claiming that customers had completed application forms incorrectly or not met the required stringent timescales for their completion. Other means of address included them offering customers flights from airports up to 200 miles away in the hope that this would stop them redeeming the offer. Other 'dirty tricks' involved them sending out applications on Christmas Eve knowing that mail would be delayed, and their customers would miss the timescale for returning them. If this tactic did not work their next port of call was to claim that their applications had gone missing.

There was controversial media coverage around this time that not one single flight had been issued and thousands of their customers, angry about the way that had been treated and misled, began to take action against the organisation.

By 1993, Hoover was making £24m in losses and the company was imploding with senior executives blaming each other for the debacle. Despite being partially bailed out by their parent company, namely Hoover Europe, it was irreparably damaged in terms of its reputation and financial standing. Their market share plummeted which was compounded by a flurry of second-hand Hoover vacuum cleaners flooding the markets by those who had taken up the offer. Front page news headlines at the time read 'Cleaning up after Hoover' and 'More turbulence for free flights'.

More than three decades on the blunder is widely regarded as the worst sales promotions in history and a lesson of the reputational risks of what can occur when a company falls short on a customer promise.

2.11 Stakeholder Value

Stakeholder value can be described as the adoption of management principles which seek to create value not just for business owners but for all other parties who may have interests in the business. These include customers, employees, the supply chain, partners, regulators, investors, the local community and the wider public at large. Companies sometimes create stakeholder value propositions (SVP's) which encapsulate clear and concise statements which summarise the value that companies provide for all their stakeholders. These SVPs should articulate the outcomes and benefits that a company and its stakeholders expect from one another. Stakeholders are very important to organisations as without them there would simply be no projects. Engaging with them especially around decision-making processes can influence the actions and behaviours of an organisation and greatly assist project management teams. Whilst employed in the public sector undertaking major capital redeployment and regenerative for colleges and universities the books co-author Professor Challender has found that the involvement of certain stakeholders is essential for project success. This has included local authorities who have a vested interest in making sure that the types of development in the local area are conducive to their own aspirations in terms of the mix of development, employment potential, providing local housing and any planning restrictions. In a similar way having the involvement of public funding agencies and bodies, alongside private investors are crucial to projects going ahead where financing is a key consideration. Such investors and funders will want to be assured that their money is being spent wisely, and the right decision-making is being upheld. From time to time, they may wish to influence the decision-making process and assess with project teams the various alternative propositions and the respective benefits and pros and cons of each.

Whilst closely intertwined with broader financial and economic value, shareholder value is distinctly different in certain dimensions. Most notably, shareholder value has at its core, achieving a return on investment for the shareholders alone. This is imperative as shareholders are the ones who invest their own funds (or funds on behalf of others) to help organisations grow. In turn (in theory) the organisations will lead to increased recruitment and greater financial returns (and therefore greater contribution to governments through taxes paid). However, this may not always be the case. Shareholder value is a key ingredient in organisational success, but it should not be the sole measure by which organisations are judged. Whilst shareholder value aligns with other types of value such as financial and economical, it is different as it has clear intended recipients. This can therefore be open to abuse as a focus on shareholder value alone can be at the detriment to an organisation's economic value and financial health. There are many examples where shareholder value has been focused upon, yet the organisations financial health has been neglected. One of the most prominent examples within the UK construction industry is that of Carillion. For many years Carillion returned 'healthy' dividends for shareholders and illustrated a high level of shareholder value. Yet after the organisations collapse, this shareholder value creation was revealed to be one of the only types of value Carillion had created, with the organisation itself in very poor financial health (which led to its collapse) and had been illustrating poor leadership and supply chain value practices for many years.

2.12 Ethical Value

According to Auerswald (2009) ethical value has in the past been attributed to those companies who use some of their profits for non-financial purposes or alternatively adopt ethical practices which could ultimately increase their production costs. One example in the UK could be the Body Shop who invest some of their profits, commonly referred to as *ethical residual*, in conservation and the protection and welfare of animals. Auerswald (2009) described the existence of an ethical residual as the medium which makes it possible for products that have a clear and persuasive ethically based brand to be priced at a premium. In addition, the Body Shop has been synonymous with a movement to guarantee that none of its products have been tested on animals. This ethical stance has no doubt cost the organisation more in research and development, thereby increasing their overall manufacturing costs, some of which would have to be passed on to customers in terms of higher prices for their products. This would in normal circumstances put the business on a lesser competitive standing against its competitors. Notwithstanding this potential disadvantage the business has attracted new customers who support its ethical credentials and therein prepared to pay more for their merchandise. This is where ethical value can in fact spill over to create financial value as what can only be described as a virtuous circle. Other examples of this model around the world. One example in the USA is Ben and Jerry's ice cream, where a vast proportion of profits are redirected into a charitable foundation aimed at supporting the world's most vulnerable people. Again, they have found that their ethical value has been bolstered by such social entrepreneurship and customers are clearly satisfied and prepared to pay more for their ice cream products than their competitors.

2.12.1 Definition of Ethics

Challender (2022) articulated that ethics has been described in the past in many ways

> ...the philosophy of human conduct with an emphasis on moral questions of right and wrong' (Helgadottir, 2008), 'the system of moral values by which the rights and wrongs of behaviour are judged' (Rosenthal and Rosnow, 1991) and 'a moral philosophy that involves systematizing, defending and recommending concepts of right and wrong behaviour.

Alternatively, ethics could also be defined as:

> ...the systematic attempt to make sense of individual, group, organizational, professional, social, market and global moral experience in such a way to determine the desirable, prioritized ends that are worth pursuing, the right rules and obligations that ought to govern human conduct, the virtuous intentions and character traits that deserve development in life and to act accordingly (Petrick and Quinn, 2008, as cited in Challender, 2022)

A general definition of ethics was provided around the philosophy of human conduct with an emphasis on moral questions of right and wrong.

2.12.2 A Brief History of Ethics and Theoretical Perspective on Ethics

According to Challender (2022), the known history of pure ethics or moral (ethical) theories began with some of the ancient Greek philosophers such as Socrates, Plato and Aristotle. After recovery by English Positivists, ethics were the main area for discussions during medieval times in Europe. In more modern times throughout the 19th century ethical ideas and theories were debated robustly throughout Europe. Evolution concepts emerged into physical sciences as well as the development of ethics during this time and were supported by the likes of Darwen, Comte and Spencer. Some of the main early philosophers as provided and briefly explained by Challender (2022) are included below:

1. **Socrates (5th Century BC)**
   Socrates was the founder of the Science of Ethics and professed the moral beliefs and virtues related to controlling one passion to reach wisdom and refraining from vice and evil.
2. **Plato (5th _ 4th Century BC)**
   Plato defined the social good and their relationships in his famous book of the Republic. He advocated the four virtues belonging to the state which were justice, courage or fortitude and temperance.
3. **Aristotle (4th Century BC)**
   Aristotle advocated that people should balance emotion and rationality and ethics should be predicated on moral choices and reasoned guiding actions with objectivism at its heart.
4. **Early Christianity**
   Early Christianity focused on having a sense of one's own personal goodness and morality and professed that this was the foundation of ethics.

Ethical values should create the moral compass for which individuals live, play and work. In this sense they should present a pathway for which to formulate ideas and plans alongside future decision-making. The presence of ethical values in an organisational context provides the framework for creating a shared ethical culture. Within this environment staff should be made to feel empowered and motivated to not simply adhere to established rules and regulations but adopt ethical practices by doing the right thing. Sometimes ethical value has been purported to be something for the common good through the actions and practices of societies.

It is normal especially in large organisations to have established value-based codes of ethics that all employees should uphold. This may be related to behaviours and respect for others or could be associated with corporate social responsibility, environmental sustainability or health and safety of the workforce. Sometimes ethical values such as those referred to can be adopted by professional bodies as well as organisations. An example could be the Royal Institution of chartered Surveyors (RICS) who have created an ethical code of conduct by which all its members should operate within. These codes of conduct govern the way that chartered surveyors behave. Members must sign up to adherence to these codes and therein become accountable for any infringement. There can be severe consequences and disciplinary action for breaches which in serious cases could lead to removal from the institution or even face criminal proceedings.

2.12.3 The Challenge and Dilemma for Achieving Ethical Value in the Construction Industry

Construction related ethical dilemmas can emanate from a wide and different range of issues that may be to do with planning, conservation, waste management, land contamination, water, irrigation, pollution, energy and many more issues. They can differ by nature, scale and effect but it seems they all have common dimensions within the context of ethics. Accordingly, it is vitally important that organisations and companies demonstrate their commitment to ethical and business principles.

The space shuttle Challenger was tragically destroyed in 1986 during its launch, despite warnings by engineers, which were overridden by managers. According to Uff (2003) '*...the debate continues as to whether the warnings were sufficient and whether the engineers should have prevented the fatal launch*'. More recent cases including the Deepwater Horizon oil rig explosion in 2010 and the Grenfell Tower fire in 2017, where 72 lives were lost, are testament to whether ethical lessons can be learnt to safeguard against further tragedies of this kind.

The issues around ethics affect a large spectrum of the population including private client organisations, local authorities, consultants, suppliers and contractors. Construction professionals make judgements and use their own skills and knowledge as to their behaviours and the way they operate. They are ultimately responsible to their clients under their institution's professional code of ethics. Almost every profession globally has a code of ethics which is intended to provide a framework for establishing the correct and proper ethical choices. Notwithstanding this premise, sometimes such professional ethics are not always aligned with business ethics, which are in the latter case predicated on making profits and meeting commercial targets of their respective organisations. In such cases, financial considerations can become a pertinent factor in an imperfect environment where competing factors can create ethical dilemmas. According to FMI (2006) this can be exacerbated by fierce competition to win projects, low price mentalities and small contractor profit and overhead margins. Clearly this can create challenges and if not managed and regulated effectively, repercussions for the reputation of the construction industry.

According to Parson (2005), there is a large discrepancy between the value people place on ethics and what they do in practice to support their values. The common phraseology to describe this phenomenon would be that such people *'do not practice what they preach'.* What reinforces this prognosis is that whilst a large majority of construction related individuals believe that there need to be an industry-wide code of ethics, only a minority of them would agree that adding regulatory measures to enforce ethical behaviours is something they would want to introduce. Another substantiation for the same argument is that few construction companies still do not include ethical issues as part of their strategic plans or mission statements. Furthermore, some organisations have no ethical guidelines, policies or programmes for their staff to abide by. In such scenarios, it is perhaps not surprising that ethical considerations play 'second fiddle' to commercial matters and ethical breaches are commonplace in the construction industry dominated by profit margins.

Clearly this ethical dilemma needs to become 'centre stage' in the priority stakes for construction firms to take it seriously. Possible improvement measures and reforms could come from business leaders in setting thresholds within their respective organisations for personal ethical standards which could be reflected and applied down the line to their employees. Compliance with such ethical standards may be something that could be enforced through human resource management (Parson, 2005). This could take the form of training and education designed to make employees aware of what is acceptable and unacceptable practice and behaviours in the workplace. Implementation of measures and disciplinary procedures would need to be in place as a deterrent for breaches and to encourage good practice. However, this is easier said than done in practice. According to Parson (2005) levels of warnings for unethical behaviours would need to be managed together with disciplinary proceedings in certain cases. Processes to deter bad practices should also be in place. One example of this could be safeguards for financial and procurement processes to prevent impropriety and false accounting practices.

2.12.4 Ethical Principles and Values

It seems that ethical and moral principles are not as clear as black and white when considering whether certain behaviours are good or bad or a view on whether business dealings are right or wrong. For instance, some may refer to an increasing drive for increased profits as human nature, but others could perceive this as greed. The overriding consideration is perhaps not profiting per-say but the manner and morality they deploy in pursuit of such objectives. In an ever-increasing competitive world, there could be a misguided focus on self-interests, increasing market superiority and the battle for power. This became more important during the global Coronavirus pandemic outbreak in 2020 where some construction related organisations are struggling to survive and others seeking to take commercial advantage of an international crisis. The fallout from the Coronavirus pandemic and the ethical and unethical practices that emerged from these present good and bad examples of practice and behaviours at a time of global uncertainty. In such instances, on both an individual and organisational level, this could result in cutting corners, disregarding ethical values and principles and failing to recognise the health, safety and well-being of subordinates, stakeholders and communities.

Considering the above concerns, it is imperative that professionals working within the construction industry are to consider ethical principles and human values in their daily lives. Within the context of the Law, they must work in compliance with legal and statutory requirements. Within the context of professionalism, they must adhere and follow their professional institutions' codes of conducts. Within the context of humanity, they should and have to apply their principles, ethics and values as human beings. Examples to reflect the applications of above within the framework of the Law are matters relating to health and safety, planning and environmental infringements. In the latter case, the job of environmental ethics is to outline our moral obligations in the face of such concerns (IEP, 2010). On a purely professional level, upholding the qualities, standards and codes of conduct of one's own institution and taking decisions which can have long lasting effects on stakeholders, communities and possibly humanity are all important ethical considerations. Other considerations include avoiding conflict of interests, safeguarding sensitive and confidential information, always acting with integrity and honesty, gaining trust and respect of others and maintaining equality, diversity and inclusion human resources policies. In addition, on a personal level; improving the quality of one's own practice and competency through reflection and evaluation of one's own practice, continuing professional development (CPD) and completing personal development plans (PDP) are all prerequisites of good ethical practice.

Professional conduct and ethical duties of individuals should be applied in the wider public interests and not only confined to the interests of clients and employers (Uff, 2003). Professionals, who are working in the construction industry, and often need to take actions and make decisions, must not only consider the interests of their clients and stakeholders, but the wider communities and the environment as well. On this point, Fewings (2009) argued that personal values and ethics influence judgement on such actions and decision-making. These in turn may have significant impacts and consequences on health, safety, social and lifestyle issues and consequences can affect local, national or international communities.

2.13 Sustainable Value

Sustainable value is related to the concept of integration. It seeks to integrate the environmental, economic and social dimensions of sustainability into investment decision-making and financial value. In this sense sustainability can be defined as satisfying the demands of the present without compromising the potential needs of future generations (United Nations Brundtland Commission, 1987). Sustainability has also been described as the ability of society to exist and develop without depletion of the natural resources required to live in the future. Furthermore, Kruger and Crowne (2019) advocated that there are the 4C's of sustainability and these are commerce, community, conservation and culture. Examples of sustainability include the use of renewable technologies for energy production including solar and wind, and sustainable construction. There can be many benefits for organisations in adopting sustainable practices including reducing costs associated with energy usage, waste and maintenance whilst increasing operational efficiencies. For organisations having core values linked to social consciousness and sustainability are ethically responsible. This can have benefits associated with raising their reputational value as referred to earlier in the chapter and therefore be considered a good investment option and attract the best staff. Accordingly, sustainable business strategies are those which organisations create and implement with a view on creating improvements on the environment and their impacts on communities. In addition, sustainable value creation (SVC) is the means by which to create sustainable value for all stakeholders through rethinking systems and transforming existing business models. It offers organisations both financial benefits and those linked to social value and the environment.

2.14 Well-being Value

This refers to the overall benefit to an individuals' physical, mental, emotional and social health resulting from an initiative, service, or policy. It extends beyond financial outcomes to encompass how people feel and function in their daily lives. It can be focused internally (on staff) or externally (on clients, customers and members of the public). Well-being value is increasingly recognised as a vital component of sustainable development, organisational success and general social progress. When individuals thrive, communities and organisations are more likely to prosper. This can even be extended to positive local, national and international outcomes. Well-being value includes the measurable and perceived improvements in satisfaction, morale, productivity and health. For example, a company offering flexible working hours, mental health support, or meaningful work can create positive well-being outcomes for internal and stakeholders, which in turn enhances performance and reduces absenteeism and turnover. Organisations can apply well-being value by embedding it into strategic thinking at a high level, but this must be reinforced with actionable steps and delivery of planned initiatives. Having in place a supportive workplace culture will not only aid the delivery of internal well-being initiatives but also support in the delivery of externally focused well-being initiatives.

2.15 Religious Value

Arguably some of the oldest values are those that originate from religion. The moral and ethical principles and resulting guiding actions that are derived from religious teachings cover some of the most fundamental behavioural values. These include honesty, compassion, respect, humility and being supportive of others and serve to guide wider societal support systems and underpin all actions undertaken. There are countless examples of figures throughout history and across all religions who have undertaken ethical organisational practices derived from their religious beliefs that have gone above and beyond legal or societally expected behaviours. As such, religious values can serve (and have served) as a strong foundation for shaping organisational culture and guiding decision-making. In contemporary business practices, whilst religious values may not visibly be at the forefront of organisational values, they may be closely intertwined with ownership intentions and management practices that will then in turn underpin the organisational actions.

2.16 Freewill Value

This is an often-overlooked value, and one that the citizens of many countries take for granted. Freewill value refers to the ability of individuals to act with autonomy and free from persecution from authorities for their everyday actions. Obviously, a clear line has to be drawn around what are and are not acceptable behaviours, and in many cases, legislation is the line that must not be crossed. However, in some countries (past and present) legislation has been a tool abused by those in power to remove the freewill of certain citizens. In other cases, there may be legislation in place, but it has simply been ignored or unenforced sufficiently, and this too has resulted in the removal of freewill value from certain citizens. The ability to make choices and live a free and fair life is a basic human right that, whilst afforded to many across the globe, is unfortunately not afforded to many others. It may not seem immediately apparent how some public and private sector organisations could create freewill value, but in many cases they do. If they feel legislation does not go far enough in supporting rights, then the organisations can go beyond the legislation. If the organisations disagree with an approach adopted by governments, then they have the power to exercise their voice and support positive changes in legislation and approaches, to create and not hinder the freewill of their employees, and by extension, of broader citizens.

2.17 Leadership Value

This refers to simply setting a good example. As parents seek to set good examples of behaviours for their children (in most cases, hopefully), organisational leadership should seek to set a good example for their employees, for other organisations and for wider society. Leadership value is having in place the strategies and actions that allow organisations to make the positive changes to society that they intend to. Some organisations lead by example. They do good and share the best practice widely. This is sometimes viewed as self-promotion, and whilst not entirely an incorrect view to hold, self-promotion is effective and acceptable when it is underpinned by clear leadership ability and examples. Both internal and external communications should highlight an organisation's clear leadership value. If there is not enough evidence from an organisation for such communications to be undertaken, then the organisation may not be able to consider a leader in value creation. This is not uncommon, as many organisations are still creating and encouraging the creation of value, but organisations should strive to be leaders of value creation and clear in their strategic journeys regarding the type of value they create and how it will be created and shared.

2.18 The Interconnection of Value Types

Whilst this chapter has highlighted many of the differing types of value that individuals and organisations can create, an emphasis needs to be placed on the process of value creation not being undertaken in a silo. This is both from the perspective of the individuals involved in the value creation and from the types of value created. For the involvement of individuals and organisations, it needs to be understood that value is rarely created alone. Initiatives should not even be designed in isolation. Collaborative and co-creative practices need to be embraced to ensure that all strategies and planned initiatives are as inclusive and as effective as possible. Through this process, the numerous different types of value created or enhanced will become evident. The types of value created can (and will often) be mixed and multiple. One type of value is rarely created in isolation from at least a few other types. For example, if an organisation focuses on achieving ethical value in its practices, this could then resonate with customers, which in turn will create financial value. Ensuring cultural value is achieved for the internal workforce of an organisation will positively influence well-being value. Delivering social value, if communicated effectively, will lead to reputational value, and strength across all these areas will lead to a clear pattern of leadership value. Essentially, whilst the type of 'primary' value an organisation is aiming to create should be identified, the value type will not be created and delivered in isolation. An increased awareness of all the different value types will better prepare an organisation to maximise the overall value creation of its initiatives.

2.19 Summary of Chapter

The concept of value is one which has been misinterpreted and commonly overused in the past, which is not surprising given its subjective nature. Although it has been traditionally linked to the project owners' objectives in the past, which have been mostly associated with cost, time and quality parameters. However, it is increasingly being used across a broader spectrum to represent other internal and external stakeholders' visions, requirements and needs on projects. Owing to its subjective nature, value is commonly misunderstood and can vary between different individuals. For this reason, the different internal and external stakeholders including clients, end users, design consultants, contractors and subcontractors will see value from many different viewpoints and spectrums.

There are many different categories of value, and these are summarised below.

1. **Financial Value**
   The financial value of something is the price which a willing purchaser would be prepared to pay to a willing seller for something in an unregulated market. In terms of accounting, financial value represents the monetary worth of something which could be an asset. Furthermore, the concept of value from an economic perspective represents the cumulative sum of all the rights and benefits resulting from ownership. Sometimes financial value can be linked and associated with net worth, which is essentially the difference between what people own and what they owe. However, for a business, it could be very dependent on other factors such as future business, predicted market volatility, predicted sales and revenue, depreciation costs of its assets and potential future capital replacement liabilities.
2. **Corporate Value**
   Corporate value refers to the core principles which guide the behaviours, culture and decision-making within an organisation. Corporate value extends far beyond financial performance as it encapsulates moral standards and principles that the company serves to uphold, which can serve to provide the right pathway for a company to proceed along. For a business, corporate value could be demonstrated by a commitment to all other things non-financial.
3. **Employee Value**
   Employee value and employee value proposition (EVP) have been described as the statement of values, support, company culture, recognition and rewards that employers give to their employees. It is important as it can assist in differentiating an organisation in a competitive marketplace from other competitors and therein attract the most talented staff. Employee value propositions also assist employers in making the right decisions by constantly asking themselves whether decision-making is echoing the values of the organisation.

4. **Social Value**
   Social value relates to the changed experiences that people face and the outcomes and impact of these on their lives and contributes to their overall well-being. In this sense, some changes are more significant for them than others, and social value is concerned with quantifying the importance and significance that change has on people's lives. Social value is generated from the short- and long-term collective well-being of people. It is something created by social entrepreneurs that goes over and above what businesses and traditional entrepreneurs deliver. In the UK, social value emerged from the Public Services (Social Value) Act of 2012. This Act advocated that organisations should look beyond the financial value of each individual contract they undertake but alternatively focus on the collective community benefit.
5. **Cultural Value**
   Cultural values represent the core ideals and principles upon which our communities are built and represent the beliefs of a particular culture on what is right and good. Having an appreciation of which cultures people associate themselves with should allow society to better interact with individuals around them. This knowledge and awareness can encourage respect and empathy for others and assist in celebrating differences and similarities amongst different cultures.
6. **Reputational Value**
   Reputational value is frequently referred to as 'brand value' and represents the perceived value of an organisation or brand. Reputation is very important for businesses as it has a major influence on trust and loyalty predominantly for customers. It is also a necessary factor in developing partnerships and building good employee relationships. Reputation also influences the creation of the right environment in terms of a more open, productive, inclusive and enjoyable working place. It can serve as an intangible asset for an organisation and thus have an influence on the perceptions and decision-making of stakeholders. Furthermore, funders are more likely to invest in companies that have the best reputations and brand values rather than those that are perceived in a negative way. It is the reputation of a person or company that makes them special and distinctive in the marketplace. This can be the deciding factor on whether customers will be loyal and buy a particular product or not.
7. **Stakeholder Value**
   Stakeholder value can be described as the adoption of management principles which seek to create value not just for business owners but for all other parties who may have interests in the business. Stakeholders can include customers, employees, the supply chain, partners, regulators, investors, the local community and the wider public at large. They are very important to organisations as without them there would simply be no projects. Engaging with stakeholders, especially around decision-making processes, can influence the actions and behaviours of an organisation and greatly assist project management teams.
8. **Ethical Value**
   Ethical value has, in the past, been attributed to those companies who use some of their profits for non-financial purposes or alternatively adopt ethical practices which could ultimately increase their production costs. Ethical businesses such as the Body Shop in the UK have found that their ethical value has been bolstered by social entrepreneurship, and customers are clearly satisfied and prepared to pay more for their products.

9. **Sustainable Value**
   Sustainable value seeks to integrate the environmental, economic and social dimensions of sustainability into investment decision-making and financial value. It has been defined in the past as satisfying the demands of the present without compromising the potential needs of future generations. It has also been described as the ability of society to exist and develop without depletion of the natural resources. Initiatives that encompass sustainable value have included the use of renewable technologies for energy production, including solar and wind, and sustainable construction. In addition, sustainable value creation (SVC) is how sustainable value can be generated for all stakeholders through rethinking systems and transforming existing business models. It offers organisations both financial benefits and those linked to social value and the environment.
10. **Well-being Value**
    This refers to the overall benefit to an individual's physical, mental, emotional and social health resulting from an initiative, service or policy. It extends beyond financial outcomes to encompass how people feel and function in their daily lives. It can be focused internally (on staff) or externally (on clients, customers and members of the public).
11. **Religious Value**
    Arguably some of the oldest values are those that originate from religion. The moral and ethical principles and resulting guiding actions that are derived from religious teachings cover some of the most fundamental behavioural values. These include honesty, compassion, respect, humility and being supportive of others and serve to guide wider societal support systems and underpin all actions undertaken.
12. **Freewill Value**
    This is an often-overlooked value, and one that the citizens of many countries take for granted. Freewill value refers to the ability of individuals to act with autonomy and free from persecution from authorities for their everyday actions.
13. **Leadership Value**
    This refers to simply setting a good example. Like parents seek to set good examples of behaviours for their children (in most cases, hopefully), organisational leadership should seek to set a good example for their employees, for other organisations and for wider society.

References

Auerswald, P. (2009). Creating social value. *Stanford Social Innovation Review*, 7(2): 51–55.

Challender, J., Farrell, P., and McDermott, P. (2019). *Building Collaborative Trust in Construction Procurement Strategies*. Chichester: Blackwell-Wiley.

Challender, J. (2022). *Professional Ethics in the Construction and Engineering Industry*. Chichester J: Blackwell-Wiley.

Challender, J. (2023). *Organisational Ethics in the Built Environment*. Chichester: Blackwell-Wiley.

Fewings, P. (2009). *Ethics for the Built Environment*. London: Routledge.

FMI (2006). *The 2004-2005 US construction training report.* Available from: http://www.fminet.com/global/articles/treport.pdf (accessed 20 July 2019).

Fowler, T.C. (1990). *Value Analysis in Design.* CRC Press.

Helgadottir, H. (2008). The ethical dimension of project management. *International Journal of Project Management*, 26(7): 743–748.

Internet Encyclopaedia of Philosophy (IEP) (2010). *Ethics and self-deception.* Available from: http://www.iep.utm.edu (accessed 21 February 2013).

Khalife, S. and Hamzeh, F. (2020). Measuring project value: a review of current practices and relation to project success. In: I.D. Tommelein and E. Daniel (eds.). *Proceedings of the 28th Annual Conference of the International Group for Lean Construction (IGLC28)*, Berkeley, California, USA. https://doi.org/10.24928/2020/0044.

Kliniotou, M. (2004). Identifying, measuring and monitoring value during project development. *European Journal of Engineering Education*, 29(3): 367–376.

Kruger and Crowne (2019). *The 4C's of sustainability.* Available from: https://krugercowne.com/the-4cs-of-sustainability/#:~:text=Segera%20finds%20that%20balance%20between,the%20heart%20of%20the%20development (accessed 10 June 2024).

Maher, W. (1992). *The Management of College and University Archives.* Metuchen New York: Society of American Archivists in Association with Scarecrow Press.

Parson, E. (2005). *The construction industry's ethical dilemma.* Available from: http://www.ecmweb.com/mag/electric_construction_industrys_ethical/index.html (accessed 11 November 2010).

Rosenthal, R. and Rosnow, R.L. (1991). *Essentials of Behavioral Research Methods and Data Analysis*, 2nd ed. Boston: McGraw-Hill.

Thyssen, M.H., Emmitt, S., Bonke, S., and Kirk-Christoffersen, A. (2010). Facilitating client value creation in the conceptual design phase of construction projects: a workshop approach. *Architectural Engineering and Design Management*, 6(1): 18–30.

Uff, J.P. (2003). *Duties at the legal fringe: ethics in construction law.* Society of Construction Law. Available from: www.scl.org.uk (accessed 12 September 2022).

United Nations Brundtland Commission (1987). *Sustainability.* United Nations Academic Impact. Available from: https://www.un.org/en/academic-impact/sustainability#:~:text=In%201987%2C%20the%20United%20Nations,development%20needs%2C%20but%20with%20the (accessed 10 June 2024).

3

Value in the Built Environment

Insanity is doing the same thing over and over again and expecting different results.
Albert Einstein

3.1 Introduction to Chapter

The global construction has been in the past accused of being slow to change, develop and react when compared to other sectors such as the manufacturing and petrochemical industries. For this reason, such initiatives around developments in construction technologies and new advanced methods of construction have not been adopted as they were intended. This has had the effect of creating a sense of inertia in the industry where more improved outcomes are being increasingly demanded but with little sense of learning from lessons of the past. Notwithstanding this assertion, and with reference to the above quotation, there has still been expectations for the industry to grow and deliver improved project outcomes regardless of these failing. To respond to this dilemma, this chapter will discuss the context of the construction industry, some of its inherent problems, and measures to consider value as the mechanism to drive future improvements.

The chapter will start by attempting to identify what the construction industry is, what makes it bespoke and why the procurement process around construction is very unlike that of most other industries. In this regard, it will touch on the temporary nature of construction projects and why fragmentation does not always allow the time for relationships to develop and flourish.

There will be an explanation of the traditional critical success factors of cost, time and quality, commonly referred to as the 'iron triangle'. Thereafter, the chapter will discuss the size, shape, impact and workforce of the construction industry to emphasise its scale and importance to economies for employment and as a major contributor to Gross Domestic Product (GDP). Following this, it will provide a narrative on why the global construction industry has suffered in the past from negative perceptions and a generally poor image and reputation. It will explain how these predicaments have emerged and the underlying reasons and causes behind them. In this regard, the ethical challenges for the built environment will be articulated and examined, with the aid of data compiled by reports and surveys. This will then lead into a discussion around the problems the global construction industry has faced in achieving project success, particularly in developing countries.

To respond to the above dilemmas, problems and challenges, there will be an explanation of why the construction industry has started to adopt a more values focused agenda. In this regard it will discuss why there is a concerted effort in many countries of the world to move away from the traditional fixation with the price in lieu of more value-based approaches and outcomes for projects. Measures to adopt such changes in strategies will be examined with reference to government and other public sector initiatives such as those instigated by The Commission for Architecture and the Built Environment (CABE) in the UK. In addition, Private sector initiative (PFI) projects, which attempted to provide value-centric and service-led integrated approaches, will be discussed along these same lines. Finally, the chapter will identify and describe legislation, recent reforms and other governance mechanisms which have in the past been introduced in the UK to drive value and therein act as a catalyst for improvements.

3.2 What Is the Construction Industry and What Makes It Bespoke?

> The construction industry is one of the most dynamic and complex environments as it is a project-based industry within which individual projects are usually built to clients' needs and specifications.
>
> *Tabassi et al. (2011)*

With reference to the above quotation, the built environment has been defined as embracing all inputs to the provision of managed space and infrastructure for public and private use: property investment and development, design, construction and facility management (Saxon, 2005). The construction industry has emerged as a bespoke project-based industry and there are many different theories around this. As a starting point, it is important to understand what makes the construction industry different and potentially 'at odds' from most other industries. In answering this question, it is worth contemplating that the procurement process around construction is very unlike that of most other industries. Those employed are made up of mostly small teams, ranging from construction workers to design consultants, who come together on a temporary basis for the life of a project and then disband to undertake different projects. This creates fragmentation and does not always allow the time for relationships to develop and flourish, which could be in itself a contributory challenge for trust generation. It is also important to reflect upon the 'end product' and that construction projects are nearly always bespoke to clients' requirements. This 'one off' or 'made to measure' aspect does, however, create risk and uncertainty for all parties. To fully appreciate and understand this context it may be useful to compare the procurement of a new building with the purchase of a new car. When one buys a car the make and model that suits your budget will be agreed alongside any affordable optional extras that are required. One can even 'test drive' the same model to ensure that it meets the customers' expectations in terms of feel and drivability. At this stage, on ordering the vehicle, the customer will know exactly what they will receive on the due delivery date, which is normally a few weeks at most and have an agreed fixed price. In this regard, there is very little risk that the customer will not receive exactly what they have expected when they ordered for the price they have secured. As the car is made is a factory it will be standardised, and quality control is normally very good accordingly. The complete opposite scenario could be argued to prevail when a building is being procured. It normally involves a prolonged period of time for design consultants to formulate a brief with clients, progress the design development and tender the projects to construction contractors. On receipt of tenders, this is where the process probably varies most from the car purchase example. In selecting the most appropriate tender it is important to consider the quality of the tenders rather than just accept the lowest price. Such factors as reputation, track record, resources and demonstration of an understanding of the project are vitally important as the quality, cost, scope and timescale in delivering the final project are normally anything but assured. There are three elements which are widely regarded as critical success factors namely cost, time and quality, and these constitute what is commonly referred to as the 'iron triangle' which is illustrated in Figure 3.1.

There are several unknown factors in any construction process which could cause the cost of projects to increase, programmes to be delayed and the quality of builds to be compromised. This introduces two aspects which come into play around commerciality and risk and who incurs any additional costs is frequently an area where disputes arise between clients and their consultants and contractors. Furthermore, given that construction works normally incur significant amounts of money the stakes are high in terms of the final bill for clients and the level of profit attained by contractors. It is therefore perhaps not surprising, for reasons of commerciality, that parties to construction contracts have traditionally not relied on trust in dealings with each other, especially around financial construction matters.

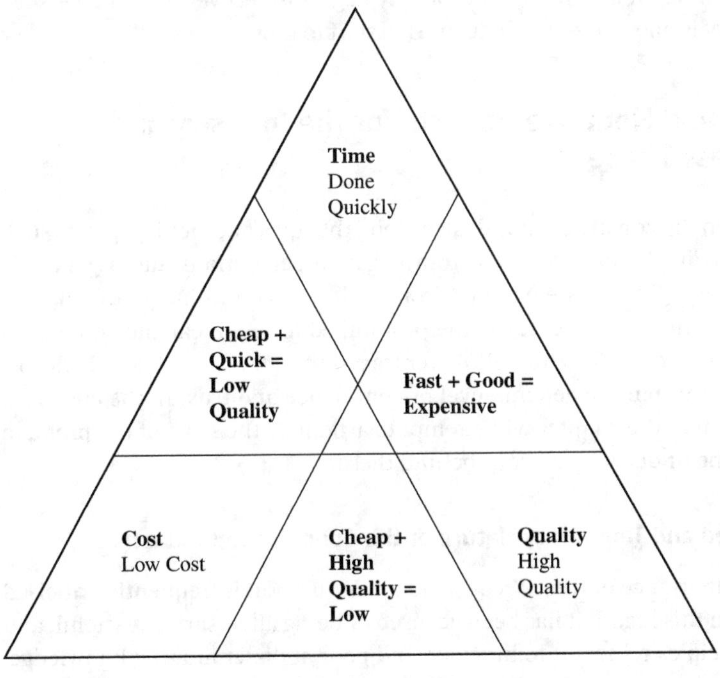

Figure 3.1 The 'Iron Triangle' of cost, time and quality.

3.3 Size, Shape, Impact and Workforce of the Construction Industry

The construction sector employs between 2 and 10% of the total workforce in most countries of the world (Rahman and Kumaraswamy, 2004). In the UK, construction output in September 2020 accounted for more than £110 billion per annum and contributed 7% of Gross Domestic Product. Approximately a quarter of construction output is public sector, and three-quarters is private sector. The construction industry is very diverse and includes elements ranging from the construction of buildings and infrastructure, maintenance, refurbishment to the manufacture and supply of building products and components. New build construction output accounts for approximately 60% of construction output whereas refurbishment and maintenance accounts for the remaining 40%. Government Construction Strategy outlines that there are three main sectors within the UK construction industry, and these are commercial and social (45%), residential (40%) and infrastructure (15%).

According to the Government's Construction 2025 Strategy (HM Government, 2013), the UK construction industry generates approximately three million jobs which accounts for approximately 10% of total employment and these include construction related services and manufacturing. These jobs can be categorised as contracting which employs approximately two million people, and made up of 234,000 businesses and services, employing 580,00 people and 30,000 businesses. Furthermore, construction related products accounts for 310,000 jobs and 18,000 businesses. The construction industry across the world can be regarded as a long-term industry linked to investment, albeit perceived to be a high cost, high risk industry by some. The performance of the industry can be seen to be a good indicator of the state and health of the sider economy. Accordingly, when an economy falters, it is common for the construction industry to slow down and when an economy begins to recover from a recession, the construction industry is one of the first industries to start to grow.

3.4 Perceptions and Negative Impacts for the Industry and Attempts to Address These

Despite efforts to develop the construction industry along the lines of other industries such as manufacturing, it has in the past suffered from negative perceptions and a generally poor image and reputation. This has emanated from a slow pace of change to embrace improvements coupled with reported cases in the past linked to unethical and sometimes illegal practices. In the latter case, the Grenfell Tower tragedy of 2017 in the UK, which cost 72 lives, has done little to unpin the general level of confidence and trust in the construction industry. This section of the chapter will attempt to articulate the scale of the problem the industry faces and the underlying reasons behind these.

3.4.1 The Fragmented and Temporary Nature of the Construction Industry

Despite the scale and importance of the UK construction industry, it is frequently criticised for performing poorly. In this regard it has been accused of being adversarial, wasteful, and dominated by single disciples reluctant to innovate and poor at disseminating knowledge. It is also claimed that the industry is fragmented and according to Fewings (2009) there is a general tendency for those employed in the construction industry to ignore best practice owing to this fragmentation within the sector. In the UK this is reflected in the fact that nearly half of all work completed is undertaken by approximately 190,000 small contractors. Some of the workforce in this regard, especially those who undertake smaller jobs are often one-off individuals with limited and potentially insufficient experience. As a consequence of such inexperience there may be cases where their knowledge and awareness of issues relating to regulations, compliance, best practice and ethical considerations is lacking. In the context of the construction industry, which is associated with fragmented, complex and potentially confrontation practices, this can create a dilemma for the sector. It is, therefore, perhaps not surprising that there have been many reported cases in the press of examples of malpractice and contractors breaches regulatory standards and codes of practice. Some of these cases have resulted in litigation proceeding brought about by their clients which has resulted in reputational damage for the industry as a whole. In addition, these practices have frequently led to less than acceptable project outcomes in terms of value for money, delays and poor-quality build standards.

Within the fragmented context of the global construction industry, teams are created as temporary organisational entities with the sole purpose of delivering projects and then subsequently disbanded at relatively short notice as projects are completed. This temporary organisational model can present a dilemma for the construction industry in it being regarded and perceived as a relatively short-term concern from a planning and delivery perspective, when compared to more operational teams e.g. maintenance and facilities departments, who oversee the occupation 'in use' post completion stages. These characteristics can adversely affect the success of projects, especially those teams assigned to project management, where there is sometimes little scope for inter-organisational relationship development and partnering philosophies. For this reason, there have been many attempts to improve the current situation and provide a more robust and stable environment for the industry. This is designed to create a more collaborative, longer-term basis for construction project management to operate within, based on the partnering principle and harbouring trusting relationships (Challender et al., 2019). Notwithstanding this premise, some project managers still adopt traditional practices and behaviours which at times could be described as adversarial and predicated on win-lose scenarios. Furthermore, in some cases, projects may be managed in an ad hoc and intuitive manner with resources being allocated to them across many different areas and stages. Clearly this could introduce unnecessary and avoidable risks by not having a more systematic and controlled basis in place, with established procedures, governance and strategies deployed. To address these inadequacies, the notion of project success needs to be more carefully considered from the outset by all stakeholders, including project managers. This is a subject matter that the book will delve into to consider why the industry has started to adopt a more values focused agenda to drive project success.

3.4.2 Ethical Challenges for the Built Environment

The issue of poor quality of construction projects and questionable professional ethics has been highlighted in the past, especially in those developing countries. With the move over recent years globally to focus on quality related outcomes rather than primarily cost and time, this has highlighted the need for construction professionals to focus on ethics.

Rahman et al. (2007) argued that the construction industry provides a perfect environment for ethical dilemmas, with its low-price mentality, fierce competition and paper-thin margins (FMI, 2004). Transparency international which represents a global coalition against corruption, found that 10% of the total global construction expenditure was lost to some form of corruption in 2004, which is staggering and further depletes the reputation of the industry. There have been growing reports over the years that unethical behaviours and practices are still common in some areas of the construction industry are still common, and this is taking a toll on the industry's reputation. FMI (2004) undertook a survey, which concluded that 63% of construction professionals felt that unethical conduct was still a problem for the industry. This clearly demonstrates that some improvement measures are required. Accordingly, the book will cover such ethical dilemmas and discuss measures, principles and models to curb such practices in the pursuit of a more regulated industry founded on moral values and behaviours. This is intended to act as a catalyst for improvements to construction management practices and improve the reputation of the industry which has suffered from reputational damage in the past. Later chapters of the book, in this way, will seek to equip practitioners with a practical guide or 'toolkit' to take business decisions which do not transgress into those areas of malpractice and unethical conduct. This is a deliberate attempt to promote best practice and regain the confidence and trust of the general public, where reputations have been damaged over previous years. Hopefully, this will deliver the benefits and impacts in terms of successful project outcomes and greater client satisfaction.

3.4.3 Inherent Problems Faced by the Construction Industry

The global construction industry has become synonymous with start and completion dates for projects. It has become dependent on project management tools rather than traditional management practices in other industries. Correct use of such tools can be beneficial if applied across the life cycle of projects rather than just considering the project design and construction stage. Accordingly, the project life cycle should represent a framework for how projects are delivered and should ensure the correct execution of all activities. It is the project managers role who act as a single point of contact to oversee how projects are conducted and who should adopt and utilise critical success factors to achieve project success. As previously referred to, the construction industry is different from other sectors as it deals mostly with bespoke projects and not merely the same products on a production line as one would find in manufacturing. For this reason, projects are unique and critical success factors identified in other industries cannot comfortably be transposed across to the construction industry (Garbharran et al., 2012).

The competitiveness in the construction industry dictates that project success is becoming more important, especially in developing countries in allowing construction organisations to maintain and attract new clients. Despite this, Garbharran et al. (2012), explained that in the context of construction projects carried out in South Africa, project failures are sadly all too common. Successfully procuring projects in developing countries has become extremely challenging, especially for those projects that are large and complex. It is for this reason that they have suggested that appropriate critical success factors as a tool for continual improvement may need to be identified and utilise in practice.

3.4.4 The Global Dilemma and Challenges in Achieving Project Success

The construction industry represents one of the largest employers across the world and is a very competitive industry. According to Garbharran et al. (2012), from their studies in South Africa, project success in the construction industry has become a rare commodity. This has caused problems in maintaining existing clients, let alone attracting new ones. Furthermore, they concurred that too much emphasis is placed on the 'iron triangle of cost, time and quality' where success is measured from projects being completed on time, within budget and to defined quality standards. Consequently, they advocated that the construction industry needs to have more focus on other success factors related to value if it is to survive the challenges presented by globalisation. Accordingly, critical success factors represent those aspects and inputs into project management systems that can increase the opportunities to achieve successful outcomes for projects and should be encouraged (Garbharran et al., 2012).

It is essential for companies to remain successful to maintain their survival within competitive business environments. The construction industry operates within a highly dynamic environment characterised by changes brought about by new technologies and business methods. This can present a dilemma for those construction related companies, large and small in being able to adapt to these ever-changing challenges and develop appropriate strategies to become more competitive (Arslan and Kivrak, 2008).

3.4.5 Risks Associated with Lowest Cost Tendering in Developing Countries

In developing economies, procurement strategies for projects have been heavily influenced by lowest price tendering and there is little emphasis on open bidding and other selection criteria other than cost. This is clearly at odds with Latham (1994), Egan (1998) and Egan (2002), who found that accepting tenders solely from a lowest cost perspective should be avoided in all cases. This assertion was predicated on a history of issues and problems from procurement strategies linked to lowest cost tendering linked to contractor claims and other project implications associated with programme delays and poor-quality delivery. Alsolami (2022) explained that this dilemma was prevalent in Saudi Arabia, where in most cases it has become mandatory in Saudi Arabia to accept bidders who submit the lowest price. It is generally not common practice for the successful contractors, appointed under these bidding processes to become part of the design team or have any input into the conceptual design stages. This therefore precludes the benefits which can be generated from the contractor's expertise and knowledge in buildability and making early contributions to design to reduce time, cost and improve quality. Furthermore, the process does not encourage collaboration, partnering and collaboration and innovation and is widely regarded as antiquated and fragmented. It also isolates the client's design team from the contractors' team, which precludes dialogue and liaison in overcoming design issues, leading to potential construction issues emerging at later stages (Alsolami, 2022). In addition, it can also have an impact on the efficiency of projects from the creation of a cultural divide between the design and contracting teams leading to a 'them and us' scenario. This can manifest itself with bad relationships developing which instead of developing win-win outcomes could lead to win-lose or even lose-lose situations.

Projects sometimes fail to meet the expectations of clients partly as a result of this very adversarial and traditional lowest tender procurement procedure which rarely introduces the best design consultants or contractors for the nature of projects. This type of procurement can result in risk issues for projects, with the design team possibly looking to transfer risk to the construction delivery team and other supply chain organisations. Accordingly, this dilemma can have severe implications for construction quality where parties are looking to reduce their own costs at the expense of the overall project. It can result in a culture of conflict, blame and claims and a reactive rather than preventative approach to problem solving.

Value management processes look to address the above failings, on an international scale linked to lowest price tendering, and introduce and enhance other selection factors into the bidding process, other than just cost. These processes seek to then evaluate design consultants and contractors' bids on measures that give added value and quality for projects, and therein avoid the failures associated with traditional bidding. It could also in this way lead to improvement in the life cycle of buildings including environmental sustainability enhancements.

3.5 Why the Industry Has Started to Adopt a More Values Focused Agenda

> Not everything that counts can be counted; and not everything that can be counted, counts.
>
> *Albert Einstein*

In consideration of the above quotation by the distinguished scientist, there is a concerted effort in many countries of the world to move away from a traditional fixation of price and cost of construction projects in lieu of more value-based approaches and outcomes. Considering value can present many challenges for the construction industry. Whilst price and cost are normally easily determinable owing to their tangible and objective nature, the same does not always apply to value. The notion of value is subjective and, as referred to in Chapter 2, can mean different things to different individuals, groups and organisations. For this reason, there needs to be a unified approach to what value means and can bring to a vast array of different stakeholders. In this way mutually aligned objectives can be determined and teams can work collaboratively to deliver them for projects as win-win scenarios for all.

In the past, the construction industry has had a traditional modus operandi for minimisation of cost on projects as opposed to optimisation of value. For all stakeholders optimising value through the delivery of other benefits other than those linked to price can present a much improved and rewarding strategy. This can be partially explained from the perspective that the initial capital costs and those associated with life cycle costs are low compared with the value added by the end user building occupants.

The Commission for Architecture and the Built Environment (CABE) was an executive non-departmental public body of the UK government, established in 1999. It produced their publication *The Value of Good Design in 2001* as a means to provide evidence to support the benefits of value-added in design from both a social and economic perspective. It was intended to act as a catalyst for changing the developing concepts and perceptions for creating value from many different stakeholder perspectives. Notwithstanding government agencies such as CABE, the construction industry has tended to approach the notion of value from a single angle in the past. This has included the perspective that value added needs to be achieved and transcended through the construction industry, largely from integrated solutions. PFI projects are one example of such integrated services, which were popular in the UK for many years, particularly in delivering large complex hospital projects.

Private Sector Initiatives drew heavily from the concept of adding value through an integrated approach, as they appeared to offer a 'turnkey' service led solutions to facilities across their life spans rather than being just concerned with the procurement of buildings. This was professed to offer clients longer-term benefits in meeting their objectives and servicing their customers. Such initiatives were designed to consider broader trends linked to the outsourcing of capabilities related to asset management. Private Finance Initiatives in the UK have since their introduction suffered negative perceptions associated with committing public sector client organisations to longer term facilities costs. This has caused a 'kick back' in their usage, especially in National Health Service in the UK (NHS) where central government funding has become more heavily audited and scrutinised for value for money.

Saxon (2005) advocated a vision for the UK construction industry around a 'value centric' and 'service-led' philosophy. He opined that integrated solutions do not always comprise service providers by simply adding in services to their clients, but in some cases, taking over those functions which had been traditionally performed by their customers. In this way Saxon 2005 advocated that this approach could 'add value' for customers and society by shaping and delivering their needs. In this sense he articulated a need to include the needs of society as stakeholders in a particular development project, against a traditional model that only followed the customer interests. This wider view of meeting other stakeholders' requirements was referred to as a 'value agenda'.

In more recent years since PFI's grew unpopular, there was still an appetite to extrapolate value added approaches to what was still regarded as an antiquated construction industry. The introduction of the *Value Toolkit* from the Construction Innovation Hub represented another initiative to increase value through change. The value toolkit was developed with more than 200 different construction experts including many UK and international consulting and contracting organisations such as Arup, BAM and Kier. Since its introduction in 2022 it has assisted in providing the means to create value and introduced ways to measure it. The toolkit was derived from value-based decision-making predicated on improving outcomes from an economic, social and environmental perspective. In this way it was designed to improve the impact of the construction industry on both existing and future generations.

The value toolkit, albeit supported by the private sector, was created by the UK government as an initiative that professed to alter the way that the construction industry considers and assesses value. It has been advocated a move away from the traditional model of only considering cost and programme for build, following two of the criteria of the iron triangle. Conversely, since its introduction sought to consider the social, economic and environmental outcomes that clients and policymakers may want to implement. The toolkit was designed to enable stakeholders to undertake informed decision-making with a view to optimising the impacts of their investments. It was also created to allow stakeholders to make decisions not just purely on cost, time and quality by introducing other socioeconomic criteria and factors linked to broader outcomes.

3.6 Legislation and Other Governance Mechanisms to Drive Value

In the past there have been many attempts by government to encourage value-added approaches to construction procurement taking into consideration those benefits previously outlined. Although not exhaustive some of these are contained below.

3.6.1 Recent Reforms in the UK Construction Industry

The UK construction industry has arguably been associated, over many years, with projects which have had less than successful outcomes. This has culminated from many different reasons, and many measures and recommendations have come forward to address this ongoing dilemma and act as a catalyst for improvements. Some recommendations for best practice and obtaining more cost and time predictability on projects include measures to improve partnering procurement processes and therein encourage more successful team integration and collaborative ways of working. This is a deliberate attempt to improve construction practices, which have arguably not been delivering the impact, and benefits that were intended in terms of successful collaborative project outcomes. Various government reports have reinforced this dilemma over the years. These have included Latham (1994), *Constructing the Team*, Egan (1998), *Rethinking Construction,* and Egan (2002), *Accelerating Change.* Other examples include *Construction 2025, Industry Strategy: Government and Industry in Partnership* (HM Government, 2013) which identifies that fractious qualities are embedded in the UK construction industry. The report stresses that collaboration and trust across the entire supply chain are crucial to deliver successful projects. More recently, this dilemma has been reiterated and become a major feature of *Modernise or Die: The Farmer Review of the UK Construction Labour Market* (Farmer, 2016). Furthermore, findings from *Low Carbon Construction Final Report* (HM Government, 2010) confirm the growing need for increased collaboration and integration across the industry, especially between the supply chain and clients, in order to make greater contributions to the pursuit of efficiencies.

3.6.2 Codes of Practice and Recommendations for Change

Throughout the decades there has been much criticism directed at the UK construction industry. Perhaps, though, the most prominent report on procurement that has had some influence on current practices is 'Constructing the Team' by Sir Michael Latham (1994). In this report Latham was critical of traditional procurement and contractual routes owing largely to a lack of coordination and integration between construction and design. He explained that in normal circumstances the main objectives of clients would include good value for money, attractive appearance, no defects, completion on time, fitness, avoidance of issues with guarantees and low maintenance costs.

Furthermore, Latham (1994) described why, owing to fundamental problems of conflict between clients and contractors, these objectives have not been realised to their full extent. He suggested that a change of behaviour through the enhancement of partnering was required to increase fairness, encourage teamwork and enhance performance through clients and consultants engaging more with contractors and their supply chain

(Latham, 1994). The Egan Reports, *Rethinking Construction* and *Accelerating Change* (Egan, 1998, 2002) followed on from this, calling for integration of design and construction phases with project teams working more collaboratively. These reports cited construction projects in the USA where modularisation, standardisation and supplier/subcontractor design innovation had facilitated greater efficiencies than those procured in the UK (MacKenzie and Tuckwood, 2012).

The UK Government's Construction 2025 Strategy was intended to spearhead transformation in the sector away from traditional practices in favour of more value driven agenda. This included the creation of partnerships between the construction industry and the Government. This was introduced to deliver longer term solutions and deliver growth in the sector. The main areas that it focused on included leadership, environmental sustainability, people, technological advancement and growth. This government strategy professed to create value through reducing both the initial costs of construction and the life cycle costs of buildings in use. Furthermore, it was intended to drive faster delivery for new builds and refurbishments through reductions in construction contract durations. The other aspirations for the strategy included reducing carbon emissions by half and increasing exports of construction services, products and materials.

3.6.3 Statistical and Academic Context for Adopting a More Values Focused Agenda

The UK construction industry annual output, ending the fourth quarter of 2017, accounted for approximately £162 billion per annum. Figures published for the second quarter of 2017 showed that public non-housing outputs reflect 11% of this total construction output, down from 17% seven years earlier. This trend was possibly explained by cuts in capital funding for public sector projects as part of the Government's Comprehensive Spending Review (HM Treasury, 2010). These figures demonstrated the impact that government policy has had on levels of capital investment in the public sector, in response to national debt and austerity measures between 2008 and 2012. Consequently, it was vital for public sector organisations to achieve more efficient outcomes and best value from their procurement strategies, taking account of expenditure cuts and reduced investment funding. Given the extent of funding challenges for the public sector, it continues to be important for procurement strategies to reflect a growing need for innovation and efficiencies. Considering these challenges, it is important to focus on a more value driven approach to overall project success.

3.7 Summary of Chapter

The built environment has been defined as embracing all inputs to the provision of managed space and infrastructure for public and private use, property investment and development, design, construction and facility management. It is a bespoke industry, and different to most other industries such as manufacturing. For instance, those employed are made up of mostly small teams, ranging from construction workers to design consultants, who come together on a temporary basis for the life of a project and then disband to undertake different projects. This can cause problems for the industry through fragmentation and does not always allow the time for relationships to develop and flourish, which could be a contributory challenge for trust generation.

The construction sector represents one of the largest employers across the world employing between 2% and 10% of the total workforce in most countries. In the UK in 2020, the construction industry generated approximately three million jobs and accounted for more than £110 billion per annum, representing 7% of Gross Domestic Product (GDP). Despite its size and global significance, efforts to develop the construction industry along the lines of other industries such as manufacturing, have in the past been disappointing. The underlying reasons could be linked to a slow pace of change in embracing improvements, coupled with reported cases in the past linked to unethical and sometimes illegal practices. In the latter case this could have resulted from its low-price mentality, fierce competition and paper-thin margins. This has caused it to suffer from negative perceptions and a generally poor image and reputation.

The fixation around the world on outcomes related only to cost, time and quality, commonly referred to as the 'iron triangle', has severely hampered the global construction industry. Accordingly, there is a need to have more focus on other success factors related to value outcomes if it is to survive the challenges presented by globalisation. In developing economies, procurement strategies for projects have been heavily influenced by lowest price tendering, and there is little emphasis on open bidding and other selection criteria other than cost. This dilemma has been a particular problem in countries such as Saudi Arabia, where in most cases, it has become mandatory to accept bidders who submit the lowest price. This practice is clearly at odds with the recommendations of Latham (1994), Egan (1998) and Egan (2002), which found that accepting tenders solely from a lowest cost perspective should be avoided in all cases.

Value management processes look to address the above failings, on an international scale, and introduce and enhance other selection factors into the bidding process other than just cost. Such value led processes can also lead to improvement in the life cycle of buildings including environmental sustainability enhancements. Examples of value led initiatives have included *The Value of Good Design, introduced by* The Commission for Architecture and the Built Environment (*CABE*) in 2001. Other examples in the past have revolved around PFI projects, linked to integration of construction and facilities management services post completion. However, these have become synonymous with a long-term expensive option, certainly in the UK, and therefore unpopular.

The introduction of the *Value Toolkit* from the UK Construction Innovation Hub represented another initiative to increase value through change. The toolkit was derived from value-based decision making predicated on improving outcomes from an economic, social and environmental perspective. In this way it was designed to improve the impact of the construction industry on both existing and future generations. Since its introduction, it sought to consider the social, economic and environmental outcomes that clients and policymakers may want to implement. Other interventions in recent times include the UK Government's Construction 2025 Strategy, which was intended to spearhead transformation in the sector away from traditional practices in favour of more value driven agendas.

References

Alsolami, B.M. (2022). Identifying and assessing critical success factors of value management implementation in Saudi Arabia building construction industry. *Ain Shams Engineering Journal*, 13(6): 101804.

Arslan, G. and Kivrak, S. (2008). Critical factors to company success in the construction industry. *World Academy of Science, Engineering and Technology, International Journal of Social, Behavioral, Educational, Economic, Business and Industrial Engineering*, 2(9): 997–1000.

Challender, J., Farrell, P., and McDermott, P. (2019). *Building Collaborative Trust in Construction Procurement Strategies*. Chichester: Blackwell-Wiley.

Egan, J. (1998). *Rethinking Construction. The Report of the Construction Task Force* (pp. 18–20). London: DETR.TSO.

Egan, J. (2002). *Accelerating Change. Rethinking Construction*. London: Strategic Forum for Construction.

Farmer, M. (2016). *Modernise or Die: The Farmer Review of the UK Construction Labour Market*. London: Construction Leadership Council.

Fewings, P. (2009). *Ethics for the Built Environment*. London: Routledge.

FMI (2004). *The 2004-2005 U.S. construction industry training report*. Available from: http://www.fminet.com/global/articles/treport.pdf (accessed 8 November 2019).

Garbharran, H., Govender, J., and Msani, T. (2012). Critical success factors influencing project success in the construction industry. *Acta Structilia*, 19(2): 90–108.

HM Government (2010). *Low Carbon Construction Final Report (November 2010)*. London: HM Government. Vol. 52–62, 196–199.

HM Government (2013). *Construction 2025. Industry Strategy: Government and Industry in Partnership*. London: HM Government. Vol. 23–25, 61–71.

HM Treasury (2010). *Spending review 2010*. The Stationary Office. Available from: https://www.gov.uk/government/uploads/system/uploads/attachment_data/file/203826/Spending_review_2010.pdf (accessed 3 December 2014).

Latham, M. (1994). *Constructing the Team*. London: The Stationery Office.

MacKenzie, N. and Tuckwood, B. (2012). A model to manage the water industry supply chain effectively. *Management, Procurement and Law. Proceedings of the Institution of Civil Engineers*, 165(3): 181–192.

Rahman, H.A., Karim, S.B.A., Danuri, M.S.M., Berawi, M.A., and Wen, Y.X. (2007). Does professional ethics affect construction quality? In: *Quantity Surveying International Conference*. 4–5 September 2007, Kuala Lumpur, Malaysia. P1–10.

Rahman, M.M. and Kumaraswamy, M.M. (2004). Contracting relationship trends and transitions *Journal of Management in Engineering*, 20(4): 147–161.

Saxon, R. (2005). *Be Valuable: A Guide to Creating Value in the Built Environment*. London: Constructing Excellence.

Tabassi, A., Ramli, M., Hassan, A., and Bakar, A. (2011). Effects of training and motivation practices on teamwork improvement and improvement and task efficiency: the case of construction firms. *International Journal of Project Management*, 30: 213–224.

4

Aligning with Existing Frameworks

4.1 Introduction to the Chapter

This chapter discusses what frameworks are and the broad nature of the concept before arriving at an adopted definition. What existing value frameworks look like, and how actions and strategies can be aligned with existing frameworks, is also discussed. The benefits of aligning with existing frameworks as well as the drawbacks that may be experienced are also outlined. How each framework is and can be used in the creation and delivery of the different types of values is also outlined. Finally, a Total Value framework is proposed that complements existing frameworks but adopts a more holistic approach encompassing the various facets of value delivery.

4.2 Frameworks

The term 'framework' can be perceived to mean a variety of different things. A governance structure, a support system, a method by which objectives can be achieved. In the context of this book, and of Value in general, we adopt a definition that spans all three of these ideas.

Firstly, 'Value' is the positive benefit resulting from a chosen strategy or action. This can be an action conducted as part of a 'business as usual' approach. Or it could refer to actions conducted over and above any legal requirement or societal expectation. It is important to recognise this distinction in that value is derived from both everyday action and additional actions undertaken beyond the minimum required. This, therefore, does not disadvantage organisations that create value as a core part of their standard operational practices. If we take, for example, a charity, through their everyday business, and adopting no additional practices, they will generate more value than a private sector organisation of the same size. However, the private sector organisation also creates value through its standard practices such as hiring of staff and generation of wealth that supports taxes, etc. Both types of organisations listed here can then also choose to create additional value through the choices they can make. This includes, for example, if they were to hire apprentices, base offices in the UK or in tax havens, buy local/national/international goods, embrace clean energy initiatives, where they invest their pension funds, the training and development they offer to staff, and the travel policies they adopt (amongst many others available).

When considering Value, A framework is the structured approach, often consisting of rules and requirements, that can be independently assessed and measured against a standard set of criteria, by which Value is realised and achieved. For the purposes of this chapter, a framework is an overarching structure that provides rules or guidance to assist in the achievement and realisation of set objectives. These objectives are to create or enhance one or more types of value.

As 'Total Value' has a diverse range of potential benefits that span numerous concepts, the range of applicable frameworks is equally diverse. Appropriate frameworks are discussed in relation to the type of Value they can help evidence the achievement of. Where frameworks span multiple Value types, this is acknowledged, but the framework is not duplicated and is only discussed once in relation to any one particular type of value.

4.3 The Sustainable Development Goals

Released in 1987, the Brundtland report defined Sustainable Development as *'development that meets the needs of the present without compromising the ability of future generations to meet their own needs'* (WCED, 1987). It has been debated how the intentionally ambiguous nature of the terms proposed in the report helped facilitate the broader acceptance of sustainability across a wide and diverse audience (Wackernagel and Rees, 1996). Indeed, the launch of the Brundtland report brought to the forefront of societal consciousness the concept of sustainability, with the terminology entering mainstream discourse. This can arguably be judged as somewhat successful, as over the prevailing decades, the concept of sustainability has been increasing in both general awareness amongst all manner of stakeholders and increasing in importance to a concept that is now extensively discussed amongst all political agendas by policy makers and decision-makers (Halkos and Gkampoura, 2021). It can be argued that within the umbrella of this ambiguity sit the Sustainable Development Goals (SDGs).

The Millennium Development Goals (MDGs) were the first globally sustainable focused international set of objectives aimed at creating a consensus of movement and action towards addressing pressing concerns of global significance. The MDGs were launched in 2000 with a 15-year time frame and consisted of eight individual goals. The aim was to start a global conversation around topics including gender inequality, disease, hunger, poverty, and addressing environmental impacts (Sachs, 2012). The SDGs replaced the MDGs when they were launched by the United Nations at their Conference on Sustainable Development held in Rio de Janeiro in 2012. The SDGs are driven by quantitative data and seek to introduce tangible targets and metrics by which progress can be monitored and measured (Hak et al., 2016). The SDGs all form part of the 2030 Agenda for Sustainable Development that was adopted by each UN Member State when introduced. The SDGs serve as an urgent call to arms for all countries to collaborate on methods, strategies, and initiatives by which the targets set for each of the 17 goals can be achieved (UN, 2025). In total, there are 17 SDGs, with a total of 159 targets and 330 indicators. The SDGs are summarised in Figure 4.1.

Despite the qualitative nature of the SDGs, there exists an element of ambiguity regarding the concept of sustainability. Research has shown that this ambiguity can lead to different stakeholders gaining different perceptions of the concept, allowing for such perceptions

Sustainable Development Goals		
1. No Poverty	2. Zero Hunger	3. Good Health and Well-being
4. Quality Education	5. Gender Equality	6. Clean Water and Sanitation
7. Affordable and Clean Energy	8. Decent Work and Economic Growth	9. Industry Innovation and Infrastructure
10. Reduced Inequalities	11. Sustainable Cities and Communities	12. Responsible Consumption and Production
13. Climate Action	14. Life Below Water	15. Life on Land
16. Peace, Justice, and Strong Institutions	17. Partnerships for the Goals	

Figure 4.1 Sustainable development goals. *Source:* Adapted from UN (2025).

to change over time, and for stakeholders to have the freedom to interpret the concept in a multitude of ways, and then use this interpretation to best suit their own business needs (Watts et al., 2019). The SDGs, however, arguably attempt to offer focus within an ambiguous concept to encourage shared action towards the achievement of each goal.

Advantages of the SDGs include:

- International Significance: the SDGs are designed to cover a broad spectrum of important element themes that both address some of the main areas of global consciousness and resonate with broad audiences simultaneously.
- International Recognition: due to being launched by the UN (an international body) and with the goals of importance to each country and different national government agendas, the goals are viewed as individually important and easily recognised.
- Qualitative and Quantitative Nature: the headline goals and accompanying metrics make them ideally suitable to those stakeholders who want to drive either quantitative or quantitative initiatives (or both).
- Action Oriented: they require action and make no apology for being focused on how each of the targets can be achieved. This can help drive forward positive behaviour change as stakeholders can be held to account for any promised progress.
- Encourage Collaboration: this is both between the goals themselves and between those stakeholders trying to achieve the goals. The SDGs try and focus on how collaboration is important in achieving the targets set.

Disadvantages of the SDGs include:

- Perhaps Overly Ambitious: whilst ambition is a good thing, it can also set too high an expectation to be achieved within a particularly tight time frame, and could then lead to all the SDGs listed being viewed as unattainable
- Apathy: due to the SDGs broad nature, this could lead to apathy amongst some stakeholder groups who see the SDGs as too broad and therefore unimportant in their day-to-day lives
- Lack of Enforceability: the SDGs are no more than advisory targets. They are relevant and significant, but the UN does not have the ability to enforce them and mandate targets on the public and private bodies needed to take action.

Table 4.1 SDGs and value type examples.

Political Value	As they are essentially politically motivated and managed, alignment with the SDGs can be seen as fostering political goodwill and helping deliver political value. This may be indirect, as through achieving (for example) well-being value created via initiatives delivered, the value will in turn aid the UK in achieving their SDG commitments. This will then allow the UK Government to gain political value in their discussion with the UN.
Sustainable Value	By their very name, the SDGs are focused on the creation and delivery of sustainable value. They are aimed at creating a better global future, and the sustainable use of current resources is a key driver in achieving this.
Ethical Value	By working towards achieving the SDGs 'correctly' i.e., in an ethical manner, then ethical value can be created by the very action of attempting to help support the achievement of the goals. It is the actions and intentions themselves that can create the ethical value in their delivery.

- Disparity of Resources: each SDG is applicable and relevant to different contexts, with the achievement of some goals more important to some stakeholders than the achievement of others. This then may not align with the resources available to each nation. Some nations may not see the significance of some SDGs, yet have the resources available to address them. Other nations may be suffering from the impacts of the areas each SDG seeks to address, yet not have the resources available to address.
- Global not Local: whist the SDGs do have to adopt a global approach in order to be globally relevant. We now live in a more hyperlocal age, where local factors account for more than global factors in many situations. This can lead to the SDGs losing relevance when compared to locally identified needs.

As a framework, the SDGs provide a robust method in which organisations can map themselves against internationally acceptable targets. Aligning to such a framework, whilst not always applicable to local contexts, does allow for legitimacy in evidencing, as an organisation, that there is a commitment to achieving sustainable value. As the SDGs are quite broad in places regarding the activities that can be carried out under their name, they touch upon numerous types of value. Some of the types of value the SDGs can help support delivering are outlined in Table 4.1.

4.4 The UK Government Social Value Model (PPN002)

Whilst many social value framework mechanisms exist, they all measure (or claim to measure) many different elements. Where they measure the same elements, often (extremely) contradictory values can be attributed to the same initiatives. Amongst this wealth of competing social value measurement toolkits and frameworks, there are those that can be described as 'micro' frameworks that measure individual activities, and those that take a more 'macro' approach, by governing, structuring, and categorising the social value activities. First introduced in 2020, the UK Government Social Value Model was then updated in 2025 with the PPN002 (further PPNs discussed in Section 4.6).

Theme		Policy Outcome	
Model Evaluation Question	Model Award Criteria	Model Response Guidance	Reporting Metrics

Figure 4.2 The Model Award Criteria (MAC) assessed as part of the social value Model headings.

The model helps procuring bodies outline the Social Value they want delivered as part of any procurement exercise. The headings in Figure 4.2 illustrate how Social Value is categorised and the metrics that can be used to help ensure the intended Social Value outcomes are achieved. The Model provides examples and guidance for any procuring body, regardless of experience, to maximise the specific Social Value they want achieved. It has five themes that provide guidance on the range areas where Social Value activity should be targeted:

1. Covid 19 Recovery
2. Tackling Economic Recovery
3. Fighting Climate Change
4. Equal Opportunity
5. Well-being

The advantages of the Social Value Model include:

- Flexibility: the Model can be tailored to both a local and national approach when required, depending upon the interests of the organisation involved. An international approach could also be adopted and the initiatives undertaken aligned to geographically applicable requirements.
- Adaptability: due to their intentionally broad nature as a list of headings to govern the areas of focus for organisations, the Model can cover a wide variety of existing and new initiatives, allowing organisations to more easily transition to any further requirements whilst still illustrating they are on a journey of improving their social value.
- Authority: Created and endorsed by the UK Government, the Social Value Model is not created by consultants with a motive of simply profit generation like many measurement tools and awarding marks that are currently available

The disadvantages of the Model include:

- Awareness: there is a lack of national awareness, especially outside of the public sector, which can serve to limit the growth and recognition of the Model and therefore hinder its broader acceptance and usage.
- Competition: there are several similar frameworks of excellence for social value that exist, meaning the Model is not alone and has the danger of losing recognition to others that are perhaps privately marketed.
- Broad nature: whilst providing benefits, the broad nature of the Model can also serve to undermine its intentions as it may be judged as not having enough single industry applicability to be recognised in any one industry as the leading social value framework.
- Dependency: there is a dependency of the Model for additional services and suppliers to be utilised to provide broader evidence of social value commitment. An organisation may feel the need to procure additional measures and services to evidence the requirements of the Model.

Table 4.2 The Social Value Model and value type examples.

Well-being Value	Many of the elements in the Model are aimed at increasing the well-being of internal and external stakeholders. By following the model and undertaking such activities well-being value will be created for direct and indirect recipients
Leadership Value	Adopting such a Model, especially where optional, illustrates leadership value in that the organisation arguably only adopts such a Model if they want to lead by example and are in front of all peers in regard to the proactive actioned they undertake in value achievement and delivery
Social Value	As the name of the model suggests, it is very focused on the achievement of Social Value. The Model breaks Social Value down further into smaller constituent elements. By focussing upon and achieving the smaller value components, Social Value will be achieved

The Model provides a clear framework for understanding and improving social value. It serves as organisational guidance for effective Social Value creation and delivery. As a social value framework, it provides guidance and direction for organisations as to the areas they need to focus on. Some of the types of value the Model can help support delivering are outlined in Table 4.2.

4.5 The Social Value Maturity Model

The model, first introduced by Watts (2025), is in response to the increasing importance placed upon social value in UK construction. This importance continues to increase significantly since the Public Services (Social Value) Act (2012) mandated its consideration in public sector procurement. This shift compels contractors to demonstrate social value commitments to win public work, prompting widespread use of social value quantification tools. These tools assign financial proxy values to social initiatives to enable comparisons between tenders. However, these tools vary significantly in how they value identical activities, raising concerns about their validity, consistency, and susceptibility to manipulation. Public clients need transparent, quantifiable data to meet procurement obligations, yet social value is inherently subjective and context specific. This tension has led to a reliance on reductionist quantification methods that favour future promises over actual capability. Critics argue these tools ignore a contractor's past performance and organisational capacity, creating a gap in evaluating whether organisations can genuinely deliver promised outcomes.

From a client perspective, ambiguity and lack of standardisation in procurement practices have made it challenging to assess social value accurately. This often leads clients to request financial totals or output counts (e.g. apprentices hired), yet such figures lack the nuance of organisational competence or ethical intention. From a contractor's perspective, many admit to inflating social value commitments in tenders to win work, even when delivery is unrealistic. Research found that some firms choose tools with more generous proxies, prioritising appearances over authenticity. In response, Watts (2025) proposed the Social Value Maturity Model (SVMM), a framework to assess and benchmark a contractor's experience, strategic alignment, and organisational infrastructure regarding social value. The SVMM offers a non-reductionist, capability-focused approach that complements (rather than replaces) existing quantification tools. The SVMM addresses several key industry needs. It accommodates the evolving nature of social value (e.g. shifting from charity donations to employment-focused initiatives), acknowledges best practices, and allows for flexibility while maintaining comparability. It also reduces the risk of social value being 'gamed' and adds a layer of organisational accountability to procurement decisions. Figure 4.3 shows the Social Value Maturity Model:

The advantages of the Social Value Maturity Model include:

- Filling a Critical Gap: by offering a reflective tool that assesses contractor capability rather than relying solely on promised outputs. The Model provides a much-needed companion framework to improve procurement transparency, contractor accountability, and social value outcomes in construction.
- Equality Between Contractors: the Model allows equality between contractors of differing sizes as it allows resources to be compared to the size of organisation and levels of achievement to be compared from one organisation to the next.
- Free to Use: whilst most tools charge the user, this framework is free to utilise for all stakeholders and offers a transparent method by which value can be understood
- Progression Tracking: the Model allows for any progression of an organisation's ability to be tracked against benchmarked levels so improvement can be evidenced.

The disadvantages of the Social Value Maturity Model include:

- The Opportunity for Measurement Tools: the Maturity Model does not prevent the continued use of measurement tools, and so there is still room for inconsistent measurement practices to occur.
- Potential to Greenwash: whilst the Maturity Model seeks to prevent organisations from greenwashing, the potential still exists for any organisations who wish to manipulate their performance to do so.
- It potentially adds to the confusion that surrounds social value measurement, frameworks, and performance, as it is another model that can be seen as 'competing' in an already crowded field.

Some of the types of value the Social Value Maturity Model can help support delivering are outlined in Table 4.3.

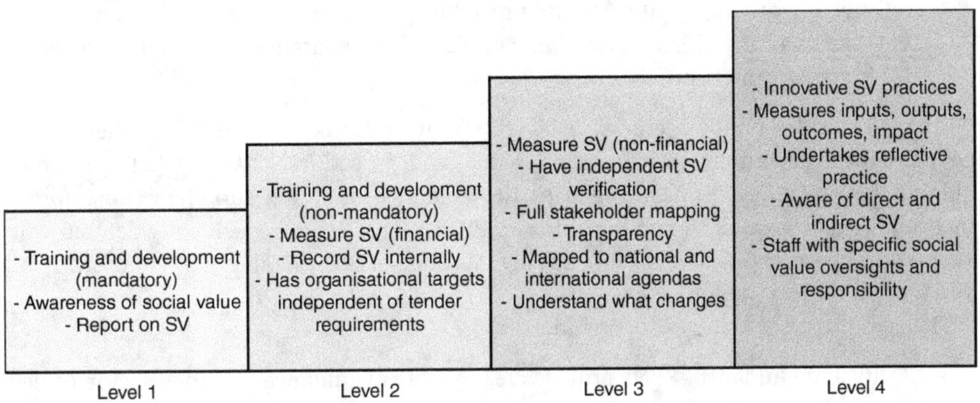

Figure 4.3 Social value maturity model. *Source:* Adapted from Watts (2025).

Table 4.3 The social value model and value type examples.

Social Value	As the Model indicates, it is primarily focused on social value. By a contractor undertaking the steps needed to achieve any of the Model levels, social value will be generated
Ethical Value	By illustrating leadership in social value, benchmarking performance, and then seeking to improve, any contractor will illustrate their strong ethical value is seeking to do the right thing for the right reasons
Well-being Value	Through the training of colleagues, an increase in team well-being can be achieved. Depending upon the training strategy at the contractor, this benefit may then be experienced by the wider contractor team as it rolls out

4.6 Public Procurement Notices

UK public procurement notices (PPNs) are official communications issued by the UK government, primarily through the Cabinet Office, to provide guidance and updates on public procurement policy. They serve as a vital tool for informing contracting authorities (such as government departments, local authorities, and public bodies) about regulatory requirements, best practices, and policy changes that affect the procurement of goods, services, and works. PPNs are not legislation themselves, but they interpret and apply relevant laws and key legal frameworks for public procurement in the UK (such as the Procurement Act 2023). PPNs help guide public sector bodies through these changes, ensuring compliance and consistency in procurement processes. Each PPN is numbered and dated, making it easy to track and reference. They typically include:

- Policy Guidance: explaining how to apply procurement regulations in practice, such as how to handle supplier selection, contract award criteria, and transparency obligations.
- Operational Advice: offering instructions for implementing new systems or tools.
- Ethical and Strategic Priorities: promoting government goals such as supporting SMEs, environmental sustainability, social value, and prompt payment practices.
- Updates on Legal or Policy Changes: addressing the impact of new legislation or case law affecting procurement rules and procedures.
- Templates and Checklists: Providing standard documentation or compliance checklists for procurement officials.

Suppliers to the public sector also benefit from understanding PPNs, as they clarify government expectations, procurement trends, and procedural changes that may affect bidding opportunities. By aligning with the guidance in PPNs, suppliers can improve their chances of success in public contracts. PPNs include those specifically dealing with Tackling Modern Day Slavery (009), Taking Account of Social Value (002) and SME and VSCE Procurement Spend Targets (001).

The advantages of using PPNs include:

- Clarity and Guidance: PPNs provide clear, practical guidance to public sector bodies, helping them interpret complex procurement regulations and apply them consistently across different departments and authorities.

Table 4.4 PPNs and value type examples.

Reputational Value	By aligning with PPNs on a voluntary basis, an organisation can enhance its reputational value and illustrate how they are embracing best practice.
Cultural Value	As the PPNs are broad and cover many aspects by embracing several PPNs, it will enhance both the internal organisational culture and the external broader societal culture.
Stakeholder Value	The PPNs cover a diverse stakeholder base, dependent upon which PPN is selected to follow. If multiple are followed, then multiple stakeholder groups will feel the positive benefit of the value based decision-making.

- Policy Alignment: the PPNs help align procurement practices with government priorities such as environmental sustainability, social value, SME engagement, and prompt payment policies.
- Rapid Response to Change: PPNs can be issued quickly to address urgent issues or policy shifts, such as during Brexit or the COVID-19 pandemic, enabling public bodies to adapt swiftly.
- Transparency and Accountability: by formalising procedures and making them publicly available, PPNs promote transparency and ensure public funds are spent in a fair and accountable way.

Disadvantages to the use of PPNs include:

- Not Legally Binding: while influential, PPNs are not legally binding. This can lead to inconsistent adoption and confusion about their status relative to formal regulations.
- Frequent Changes Can Be Burdensome: The volume and frequency of PPNs can overwhelm procurement teams, especially in smaller authorities with limited resources or expertise.
- Implementation Gaps: some public bodies may struggle to implement the guidance effectively due to lack of training, systems, or support, leading to compliance risks.
- Limited Stakeholder Input: PPNs are typically issued top-down, with limited consultation from frontline practitioners or suppliers, which can reduce their practical relevance or buy-in.

Some of the types of value the PPNs can help support delivering are outlined in Table 4.4:

4.7 B Corporation

A B Corporation (or B Corp) is a type of organisation that meets the high standards of social and environmental performance, accountability, and transparency. Unlike traditional organisations that arguably primarily focus on generating profit, B Corps are legally committed to balancing profit with purpose. They aim to create value not only for shareholders but also for workers, customers, communities, and the environment. The certification of becoming a 'B Corp' is awarded by the non-profit organisation B Lab, which evaluates organisations across several areas, including governance, worker treatment, community impact, and environmental sustainability. To qualify, an organisation must complete the B Impact Assessment, a rigorous process that measures performance against these criteria.

Only companies that achieve a minimum score and agree to periodic reassessment can maintain certification.

B Corps are often considered leaders in the values driven movement towards a more responsible and sustainable economy. They go beyond philanthropy and standard organisational responsibility programs by embedding ethical practices into the core of their operations. This can include initiatives such as reducing carbon emissions, ensuring fair labour practices, fostering workplace diversity, or supporting local suppliers. Well-known certified B Corps have built a reputation for integrating social and environmental values into their business models. Certification not only helps organisations differentiate themselves in the marketplace but also builds trust amongst consumers, employees, and investors who increasingly seek purpose-driven organisations to support. The B Corp movement reflects a growing recognition that businesses have a crucial role to play in addressing global challenges such as climate change, inequality, and resource scarcity. By holding organisations accountable to higher standards, B Corps demonstrate that profitability and positive impact can go hand in hand, redefining what it means to be successful in business.

The advantages of becoming a B Corp include:

- Enhanced Brand Reputation: achieving the certification signals a strong commitment to sustainability and ethics, boosting trust across a stakeholder groups.
- Attracting and Retaining Talent: organisations that are purpose and value driven often appeal to current and potential future employees who seek meaningful work, which in turn improves recruitment and retention.
- Market Differentiation: being a B Corp sets an organisation apart from peers in competitive industries by demonstrating verified impact, and not just marketing claims that are unsubstantiated.
- Access to a Network of Peers: Being a registered and recognised B Corp enables organisations to join a global a community of other organisations that are committed to positive change. This network access can help support and foster collaboration and growth opportunities.
- Investor Appeal: growing numbers of investors and investment bodies specifically look for certified B Corps because they demonstrate strong governance and accountability, and in theory, minimise the potential for embarrassing scandals as they are a values driven ethics organisation.

The disadvantages associated with becoming a B Corp can include:

- Certification Costs: fees for certification (which scale with organisational size) and costs of implementing improvements can be significant, especially for smaller organisations. The additional reporting and measuring commitments may also need to be covered by additionally employed staff, which would present further costs required to achieve the accreditation.
- Time-intensive Process: the B Corp Impact Assessment and ongoing reporting require substantial effort and documentation. There is no 'quick' accreditation. And whilst this is good to prevent organisations from quickly gaining accreditation when required to only lose when they no longer need, it also serves to be a drawback. Any resources dedicated to achieving the accreditation will pull resource focus from elsewhere, and so during the accreditation period, the time taken could lead to delays on existing projects.

Table 4.5 B Corp and value type examples.

Ethical Value	Becoming a B Corp illustrates operating in an ethical manner, then ethical value can be created by the very action of achieving the B Corp status.
Societal Value	Becoming a B Corp helps organisations embed social responsibility into their operations. It ensures accountability, transparency, and a commitment to positive societal impact. This fosters trust, supports ethical practices, and encourages long-term value creation for communities.
Environmental Value	By having rigorous standards on sustainability and in reducing environmental impact and emission use, being a B Corp encourages continuous improvement and transparency, promoting eco-friendly practices across supply chains.

- Legal Changes Required: some organisations must amend governance structures to legally balance profit and purpose, which can complicate decision-making and cause difficulty depending upon the existing organisational structure and its shareholders.
- Risk of Increased Scrutiny: certification invites higher expectations from stakeholders, making any missteps or failures more visible. Whilst no organisation wants to be caught in an ethical transgression or scandal, it would look much worse for those organisations that 'claim' to be of a higher moral compass, as those with B Corp accreditations do.
- Lack of Universal Recognition: whilst B Corp certification is growing, and it is increasingly recognised across the globe, it has not reached universal awareness saturation and so if an organisation does undergo all the time and expense commitments of gaining a B Corp accreditation, they then may not be recognised or appreciated for doing so by some stakeholder groups.

Some of the types of value the B Corp can help support delivering are outlined in Table 4.5.

4.8 Social Return on Investment

Social return on investment (SROI) is a framework used to measure and account for the broader social, environmental, and economic value created by an organisation, project, or initiative. Unlike traditional financial return on investment (ROI), which focuses only on monetary profit, SROI seeks to capture the positive (and sometimes negative) impacts that activities generate for stakeholders and society at large. It is especially valuable for non-profit organisations, social enterprises, and companies implementing value-based initiatives as it can help evidence the financial benefit of seemingly non-financial initiatives, by providing a way to quantify outcomes that are not easily expressed in financial terms. The process of calculating SROI involves several steps. First, the scope and stakeholders are identified. This ensures that the analysis considers those who are affected by the activity, such as employees, community members, or the environment. Next, inputs (resources invested), outputs (tangible results), and outcomes (changes experienced by stakeholders) are mapped. For example, a ready for employment training program may have the output of participants completing courses, while the outcome could be increased employment and improved quality of life. These outcomes are then assigned financial proxies, which are reasonable estimates of their economic value. Finally, the total value of benefits is compared to the investment made, resulting in a ratio. An SROI ratio of 3:1, for instance, indicates that every £1 invested generates £3 of social value. SROI reframes traditional success factors by looking at how resources create lasting social good. It helps organisations articulate the full value they bring to communities and society, reinforcing the idea that meaningful impact goes far beyond financial gain.

The advantages of becoming a SROI include:

- A Method that Helps Demonstrates Impact: SROI goes beyond financial metrics by illustrating how an initiative creates social, environmental, and economic value.
- Improves Decision-Making: the framework helps organisations allocate resources effectively by identifying which activities generate the greatest social value.
- Enhances Credibility and Accountability: It provides stakeholders, funders, and investors with transparent and structured evidence of impact.
- Supports Fundraising and Investment: a strong SROI ratio can make it easier to attract grants, donations, or impact investment by proving measurable outcomes. A high financial ratio can serve as some reassurance for stakeholders that an organisation's initiatives are effective and impactful.
- Encourages Learning and Improvement: the process highlights what works well and what doesn't regarding an organisation's strategic initiatives, allowing organisations to refine initiatives and increase the effectiveness of all strategies delivered.

The disadvantages associated with SROI can include:

- Complex and Resource Intensive: utilising SROI requires significant time, expertise, and data collection, which can be challenging for smaller and inexperienced organisations or those that do not have suitable existing skilled and experienced employees.
- Subjectivity in Valuation: assigning financial proxies to social outcomes (like well-being or community cohesion) often involves judgment calls and is therefore open to scrutiny and challenge, and any that are revealed to be inaccurate can serve to undermine all proxy values used.
- Risk of Oversimplification: reducing social impact to a single ratio may overlook qualitative aspects of value, such as dignity or empowerment. It may also force many intangible aspects to be quantified, which may lead to incorrect values being attributed to some initiatives and other initiatives being ignored completely. Such simplification may also lead users to believe that initiatives of a similar financial value are comparable when, in fact, they may be completely different and require completely different resources.
- Ongoing Measurement Burden: to remain credible, SROI must be updated regularly, demanding continuous effort and resources, which will then incur additional time and expense of the resources that need to be dedicated to maintaining SROI.
- Not Universally Understood or Accepted: like many other initiatives, SROI is not universally understood and utilised, and so some stakeholders may find the methodology too abstract or may question its assumptions, limiting its influence. Others may disagree with some proxy values or prefer the use of a different quantification system entirely.

Some of the types of value SROI can help support delivering are outlined in Table 4.6.

Table 4.6 SROI and value type examples.

Ethical Value	SROI helps organisations measure the ethical impact and ensures transparency, prioritises stakeholder voices, and supports decisions that maximise positive change. This ethical approach promotes accountability and reinforces a commitment to meaningful, responsible practices.
Well-being Value	SROI initiatives are fundamentally the well-being of internal and external stakeholders. By following the model and undertaking such activities well-being value will be created for direct and indirect recipients.

4.9 Total Value Model

A model that covers Total Value is one that essentially lists all the value types available (outlined in Chapter 2) and serves as a checklist for an organisation for the value their strategies and initiatives may create and deliver. This framework essentially serves as an educational supplement to organisational decision-making. Figure 4.4 illustrates the Total Value Model and outlines many of the main types of value to be considered.

The next step of utilising the Model Framework requires a stakeholder mapping exercise to be completed, and then the identification and listing of all actions and initiatives undertaken. This needs to be a comprehensive list of all organisational actions as Total Value is based on both the direct and indirect impacts that arise from an organisation's everyday actions as well as its specific value generating initiatives. This is illustrated in Figure 4.5.

Reputational Value	Ethical Value	Religious Value
Cultural Value	Corporate Value	Shareholder Value
Sustainable Value	Employee Value	Public Value
Stakeholder Value	Social Value	Freewill Value
Financial Value	Well-being Value	Leadership Value

Figure 4.4 Total value model value types.

Figure 4.5 Total value model.

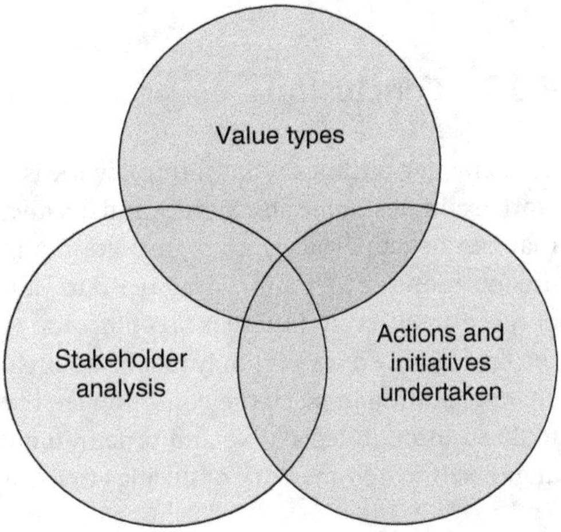

The advantages of the Total Value Model include:

- Comprehensive Nature: the model captures all the different value types that an organisation can have an impact on so that a broad and comprehensive awareness of value can be achieved.
- Flexibility in Application: the initiatives can be the ones an organisation already undertakes or can be any other initiative an organisation can perceive. The initiatives undertaken are not limited by the scope or limitations of any one tool.
- Understanding Impact: by identifying all stakeholders, the direct and indirect value-based impacts can be assessed and understood, and a true awareness of the value delivered can be arrived at.

The disadvantages of the Total Value Mode include:

- Lack of Measurement: the framework encourages measures to be put in place that best suit the stakeholders, actions and initiatives and value types. The focus is not on the quantification of value, and therefore, the model may not be utilised by those stakeholders who simply want to arrive at a total quantity of social value rather than identify and understand the true impacts and value types created.
- Approach Not Tool: the Total Value Model is fundamentally an approach to value planning. It can help with the strategic delivery and understanding of value but is not a set tool that can be depended upon instead of organisational commitment and action.
- Organisational Commitment Required: one or two individuals working in a silo on behalf of an organisation cannot undertake the requirements of the Total Value Model. It requires the commitment of many organisational colleagues from all management and operational levels and across all business units so that a full and true understanding of value can be arrived at.

As the Total Value Model includes all value types, there is no selection to be discussed as it does not prioritise one type of value over any other. Instead, it allows an organisation to understand the value it creates through its existing 'business as usual' practices. This understanding is then supplemented with an understanding of the values that are positively (and negatively) impacted by any additional actions undertaken in the delivery of initiatives purposefully aimed at value creation.

4.10 Conclusion

This chapter outlines what a framework is, before going on to provide numerous framework examples, some advantages and disadvantages of each, and some example value types that can be achieved by an organisation aligning itself with the selected framework. For organisations, leadership, with regard to value, is simply adopting best practice across one or more areas, even if there is no obligation to do so. Sometimes this best practice is even in the face of adversity, or it may just be an organisation with the resources to make a positive change, putting in place the conditions to make that positive change, without being forced to do so through legislative and procurement obligations. An organisation that decides to align itself to a framework or decides to use a framework to enhance its value delivery will ultimately adopt the role of a value leader amongst its peers. Whilst framework use in and of itself will not create value, if an organisation is serious about creating and delivering value, framework alignment can aid them on this journey and ensure the value they create is recognised, benchmarked, measured, and then ultimately, improved upon.

References

Hak, T., Janouskova, S., and Moldan, B. (2016). Sustainable development goals: a need for relevant indicators. *Ecological Indicators*, 60: 565–573.

Halkos, G. and Gkampoura, E. (2021). Where do we stand on the 17 sustainable development goals? An overview on progress. *Economic Analysis and Policy*, 70: 94–122.

Sachs, J. (2012). From millennium development goals to sustainable development goals. *The Lancet*, 379(9832): 2206–2211.

United Nations (2025). *The 17 goals*. Department of Economic and Social Affairs. Sustainable Development. Available from: https://sdgs.un.org/goals (accessed 30 November, 2024)

Wackernagel, M. and Rees, W. (1996). *Our Ecological Footprint: Reducing Human Impact on the Earth*. New Society Publishers.

Watts, G. (2025). A model of social value maturity for benchmarking construction contractors. In: C. Thompson (ed.). *Proceedings of the 41st Annual ARCOM Conference*, 1–3 September 2025. Association of Researchers in Construction Management, Dundee, UK.

Watts, G., Fernie, S., and Dainty, A. (2019). Paradox and legitimacy in construction: how CSR reports restrict CSR practice. *International Journal of Building Pathology and Adaptation*, 37(2): 231–246. https://doi.org/10.1108/IJBPA-05-2018-0037.

World Commission on Environment and Development (WCED) (1987). *Our Common Future*. New York: Oxford University Press.

5

The Theories to Help Understand Value

5.1 Introduction

This chapter explores numerous theories that can and have been applied to better explain how value can be achieved and understood. These are not prescriptive or exclusive, with more being used each week to better explore different, nuanced and innovative means of exploring the concept of value and both how it can be maximised, or why it has not been maximised in different scenarios. This chapter first explores several theories, including Legitimacy theory, Sensemaking, Stakeholder theory, Public Value theory, Critical Urban theory, Economic Urban theory and Social Constructivism, and how these theories can be used to better explore and understand the concept of total value.

5.2 Legitimacy Theory

The idea of legitimacy theory provides an explanatory lens through which any actions undertaken in the name of achieving value can be viewed. It is often utilised in a reflective and proactive manner to understand behaviours and actions undertaken. From its name, it considers how 'legitimate' any actions or strategies are and can provide a robust insight into both individual and organisational behaviour. Its underlying assumption is that actors act in an intentional manner in an effort to gain legitimacy from others. An example of this could be an organisation adopting total value practices in an effort to be perceived as more socially and environmentally active. One benefit of this would be the increasingly positive impact they now have on society. However, another would be the increased procurement success that newly adopted total value initiatives would yield. This, therefore, results in the possibility that an organisation could adopt value practices with two distinct types of legitimacy in mind:

- Symbolic – Whereby an individual or organisation doesn't amend any of its fundamental behaviours but seeks to be perceived as if it has. An example of this includes those who have been described as greenwashing. In that they haven't actually embraced any new environmental or sustainable value initiatives, but try and convince others that they have.

- Substantive – This is where an organisation or individual amends their fundamental behaviour to adopt new practices in order to focus on new ways of working that embrace a more value-focused approach.

Legitimacy theory itself, as outlined by Suchman (1995), consists of three main types, each consisting of several sub-categories, as illustrated in Table 5.1.

When considering legitimacy theory in relation to organisational value, organisations must firstly decide which sort of legitimacy they want to achieve. This may sound like an obvious question, does an organisation want to achieve symbolic or substantive legitimacy in the eyes of those they wish to be viewed as legitimate. Whilst most organisations would publicly claim they adopt value-based behaviour for substantively legitimate purposes, there are multiple examples of organisations that are publicly exposed as only symbolically engaging with value-based behaviour – almost always in an attempt to gain the legitimacy in society as an organisation that is values driven, and so receive the benefits such an association brings, but with little commitment to actually make positive and substantive change. It is therefore fortunate that the majority of organisations that pursue a symbolic values approach are ultimately exposed for doing so. The best method by which to achieve value-based legitimacy in the eyes of stakeholders is to embrace the concept substantively.

Table 5.1 Legitimacy theory categories.

Type of legitimacy	Sub-category
Pragmatic – This focusses on the practical consequences that arise from exchanges between organisations and stakeholders.	Exchange legitimacy – This is where the organisation develops practices it believes will lead to legitimacy.
	Influence legitimacy – When stakeholders believe organisations are acting for wider societal interests and therefore being responsive to stakeholder demands.
	Dispositional legitimacy – When the organisation is perceived by stakeholders as having an image associated with genuine societal concern.
Moral – This is where the stakeholder(s) perceive an organisation to be fundamentally doing the right thing.	Consequential legitimacy – Where organisations are judged by stakeholders on what they achieve.
	Procedural legitimacy – When an organisation adheres to socially accepted techniques and practices.
	Structural legitimacy – Refers to the perception of the hierarchical structure of an organisation and whether its resources can achieve its goals.
	Personal legitimacy – Evaluations are made about those personally in charge of the organisation.
Cognitive – This is when stakeholders believe an organisation's motivations mirror their own.	Comprehensibility – When the organisation purposefully structures itself in an understandable way to intended stakeholders.
	Taken-for-granted – When stakeholders view the organisation as one of the only available to deliver the service they need.

5.3 Sensemaking Theory

Sensemaking is a theory that is concerned with how individuals make sense of new or changing information they encounter. First pioneered by Weick (1995), it is a cognitive process that people follow to better understand the world around them. Whilst people are unique in their experiences and therefore interpretation of information, Weick argued that the majority undergo a similar process of understanding. This was broken down into seven key stages as illustrated in Table 5.2.

The seven stages can be followed as a consecutive process, or individuals can use one or more of the stages in a different order than what is listed. In regard to making sense of value, the sensemaking stages can be both utilised and abused by stakeholders in their communication of the value they deliver. Whilst there will be some organisations that adopt value-driven approaches without the intention of communicating more widely, for the majority of organisations, positive communication and marketing are essential elements of their value journeys. This does not always have to be considered from an ulterior motive-driven perspective, as an organisation will focus on value delivery, as it both allows them to make a positive impact on society, and it allows them to win future work. More of the latter will therefore continue to lead to more of the former.

Therefore, for sensemaking, it is a means by which value can be understood and by which value can be intentionally structured in communications. How value can be understood using sensemaking and how value can be intentionally communicated and structured is illustrated in Table 5.3.

When considering sensemaking in relation to organisational value-based initiatives and strategies, it is important for all parties to understand exactly how 'sense' can be made, and importantly, how misconceptions can be avoided. Organisations don't want to be accused of deliberately trying to mislead.

Table 5.2 Sensemaking stages.

Stage	Description
Grounded in identity construction	Future interpretations are governed by their previous experience and sense of identity.
Retrospective	Concepts are not fully understood until they have been reflected upon.
Enactive of sensible environments	The surrounding environment influences and contributes to how understandings are constructed.
Social	Meanings are essentially socially constructed and therefore only individually accepted when widely agreed upon.
Ongoing	Making sense is a continuous process as individuals are always adding to and changing their understandings.
Extracted cues	Individuals receive cues from their environment to help them make sense of information and gain context.
Driven by plausibility over accuracy	What sounds plausible can often be accepted as true.

Table 5.3 Sensemaking and understanding value.

Stage	Value understanding and communication
Grounded in identity construction	Each individual will interpret value-based communications and initiatives in a manner that is unique to them. This will be influenced by their own previous experiences and understanding of the strategy or initiative under discussion. In an attempt to ensure that value-based communications are understood as intended, organisations should ensure specific intended recipients of communications are identified and their own backgrounds (as much as can be) are acknowledged. By appreciating how communications regarding value initiatives and strategies may be understood, such communications may be tweaked to avoid any incorrect assertions being made.
Retrospective	The initial opinion individuals have about value-based initiatives and strategies may not be the opinion they hold long term. For example, individuals may be immediately impressed at the launch of a new strategy, but then, having time to reflect on it, may arrive at the view that the strategy is not as groundbreaking or innovative as they first thought. Ensuring communications are truthful and followed up with appropriate confirmation messages will aid in securing any positive first impressions, and also potentially help with addressing any previous negative impressions.
Enactive of sensible environments	This stage implies that individuals will be influenced by the surrounding environment in any understanding reached regarding a value strategy. This environment could be personal or professional and contributes to how individuals make sense of the value encountered. For example, if they work in an environment where all previous value initiatives were perceived as suspicious, then this may influence how future value initiatives are perceived. Communications regarding future value initiatives will be judged in relation to previous initiatives and experiences of the intended audiences. Therefore, having an awareness of the successes and failures of previous initiatives and the professional environment of intended recipients can help tailor initiatives and shape communication methods.
Social	This stage implies that all meanings are ultimately arrived at when shared and reinforced by others. Ultimately, the true success of value-based initiatives will be ascertained only when the initiatives have been discussed. Organisations sharing their value initiatives should be clear in their core messages and attempt to lead and stimulate conversations amongst intended audiences to avoid any confusion arising.
Ongoing	Meanings evolve and are part of an ongoing process of understanding. How individuals understand value today may not be how it is understood tomorrow. This is the very crux of this textbook, in that the concept of value has and is evolving to no longer be focused on one specific type. Any value communications need to be aware that perceptions can change, and meanings can evolve. Therefore, what was a suitable policy previously may not be a suitable policy in future. All policies need to be reflected upon and updated accordingly.

Stage	Value understanding and communication
Extracted cues	This stage links closely with an individual's environment – but as well as reaching meanings based on the understandings of others, individuals also utilise other 'cues' from their environment to help understand new and changing information, such as any value-based initiatives and strategies that are communicated. When communicating such initiatives and strategies, it is therefore important that organisations are aware of exactly how sense can and will be made of the strategies, and how existing cultural and societal cues can be used to shape how their strategies are received and understood.
Driven by plausibility over accuracy	Individuals need to remain astute regarding any claims made by organisations with regard to value achievements, as this stage of sensemaking indicates that strategies can be accepted as true and accurate if they sound true and accurate, and not necessarily because they are true and accurate. Organisations need to be aware that the data they record and report is correct. When collecting any data, this stage indicates that what sounds plausible can often be accepted as true. Organisations need to ensure all data is robustly checked and is accurate prior to using and communicating it.

5.4 Stakeholder Theory

A stakeholder has historically been described as '*any group or individual who can affect, or be affected by, the achievement of the organisations objectives*' (Freeman 1984, p. 25). Essentially, any individual with any links to an organisation. To better understand stakeholder theory, it is good to understand the traditional approach, and somewhat opposing argument, that of the shareholder theory. Whilst not developed in isolation and not strictly 'opposing' sides of the argument, stakeholder theory has been viewed as the antithesis of shareholder theory. Shareholder theory is where an organisation's actions are focused purely on profit generation to ensure shareholder returns are maximised. At an extreme, this is at the expense of other areas of focus, such as positively contributing to society or the environment. Financial performance is the singular priority, with the theory arguing, in principle, that higher profits will result in higher taxes, which will then feed into the government, which is able to, in turn, deliver more social and environmentally focused initiatives. However, this somewhat dated perspective has been replaced by that of stakeholder theory, with some even taking this theory to the extreme and arguing that organisations should exist to maximise the stakeholder benefit. When identifying stakeholders, they can be categorised as either direct or indirect stakeholders:

- Direct Stakeholders – These are individuals who are directly impacted by the decisions and actions of an organisation.
- Indirect Stakeholders – Those individuals who are still impacted by the decisions of an organisation, but in a proxy form, such as through the changes in actions of those who have been directly impacted.

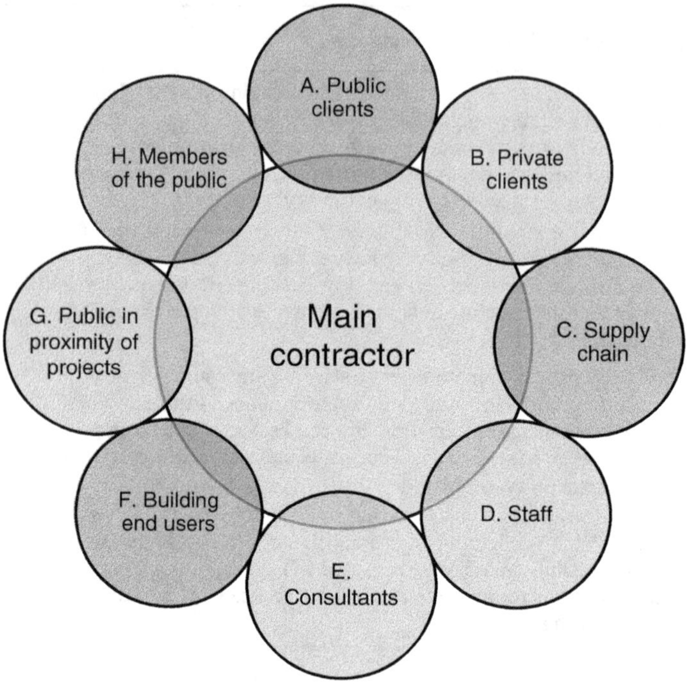

Figure 5.1 An example stakeholder map for a construction main contractor.

Organisations are often concerned with their impact on direct stakeholders and often undertake stakeholder mapping exercises to identify such stakeholders. An example of this for a typical main construction contractor can be seen in Figure 5.1.

The benefits of conducting regular stakeholder mapping exercises for an organisation include:

- Identification of each stakeholder group: it is often only through physically mapping who an organisation's stakeholders are that all stakeholders can be identified. It is best practice to undertake the mapping with several team members, and over several sessions, so the mapping can be reflected upon and updated.
- A better understanding of each stakeholder group: by clearly identifying each group, their needs can be discussed and ascertained, and the exact relationship that is in place between each stakeholder group and, in this instance, the main contractor, can be understood.
- Improved communication and relationship opportunities: by being aware of the different stakeholder groups, a greater level of clear and dedicated communication can be utilised that is tailored to the communication needs and preferences of each group.
- Enhanced risk management practices: each stakeholder group may have their own associated risks that need to be addressed in a bespoke manner. Through a better understanding of the needs of each stakeholder group, individualised risk management practices can be created that reduce and address the specific risks and challenges presented.
- Efficient resource allocation: organisational resources are finite, and so management decisions need to be made regarding the efficient allocation of resources. Some stakeholder groups may be larger than others, and some may have a larger influence than

others, or have a greater impact on your organisation. By having a greater awareness of each stakeholder, more tailored and appropriate (and hopefully successful) risk management practices can be implemented.

- Reduction of negative impact ramifications: a greater stakeholder awareness can mean that any potential negative sentiment or issues that are developing can be addressed, whilst they may remain isolated to one particular group. It may also be easier to address the stakeholder concerns if knowledge of the broader impacts is clear. By having a greater understanding of stakeholder groups and their individual needs, these can be both efficiently and effectively proactively and reactively addressed.

Whilst the above points are applicable for both direct and indirect stakeholders, all too often, it is only the direct stakeholders that are mapped by organisations. Organisations are often unaware of the impacts their actions and strategies have on indirect stakeholders. This is usually due to the organisation being unaware of who indirect stakeholders are, or having an awareness of the indirect stakeholders, but never having considered their impact upon indirect stakeholders previously.

The same stakeholder mapping exercise should be undertaken for each of the stakeholders' stakeholders. This can often be difficult with little knowledge of such stakeholders, but an attempt should still be made, as these will be the original organisation's indirect stakeholders. Such stakeholder maps can be conducted in isolation by the main contractor (in this example) or could be co-created by the main contractor and one identified direct stakeholder. The benefit of the latter is that a true and full understanding of a stakeholder's stakeholders can be ascertained. If we continue the development of Figure 5.2 by starting to map indirect stakeholders, we can map the stakeholders of the supply chain:

Figure 5.1 illustrates a direct relationship between the main contractor and the supply chain partners. Figure 5.2 illustrates a direct relationship between the sub-contractor and the materials suppliers. However, no direct relationship exists between the main contractor and the materials suppliers – but they are indirect stakeholders. By combining Figures 5.1 and 5.2, we can see that an indirect stakeholder relationship exists, as revealed in Figure 5.3.

The benefits, therefore, of conducting a robust stakeholder mapping exercise are the identification of the indirect stakeholder relationships that may be of paramount importance. This is especially pertinent when combined with total value considerations. The total value benefits of a robust stakeholder analysis include supporting a true impact map. As a theoretical lens, stakeholder theory offers a clear identification of all direct and indirect stakeholders and so allows for the potential impacts and ramifications of any strategies and actions to be understood. This is beneficial for both planned interventions, to understand the potential diverse and broad benefits they may have, and also to better understand existing strategies and actions.

- Proactive total value stakeholder mapping: this is undertaken before any specific total value action is launched and allows an organisation to fully understand the implications of its desired strategy and actions. The more detailed the mapping, the greater understanding of an initiative is achieved, and therefore the better prepared an organisation can be to both maximise and share the total value benefits achieved.
- Reactive total value stakeholder mapping: where a total value strategy or action has been launched, reactive mapping can be undertaken to better understand why it may not have been as successful as hoped, or to identify previously unknown positive impacts on further stakeholder groups. Such a technique can also be utilised to explore why certain total value initiatives may have failed with certain stakeholder groups.

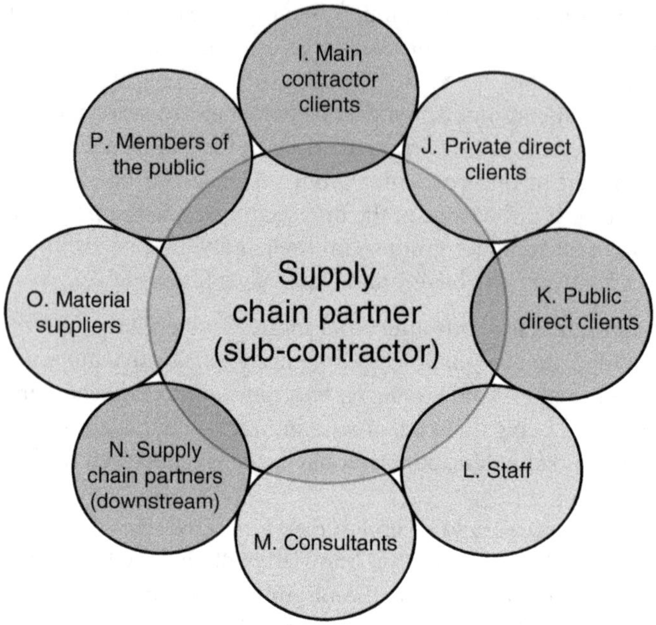

Figure 5.2 An example stakeholder map for a construction supply chain partner.

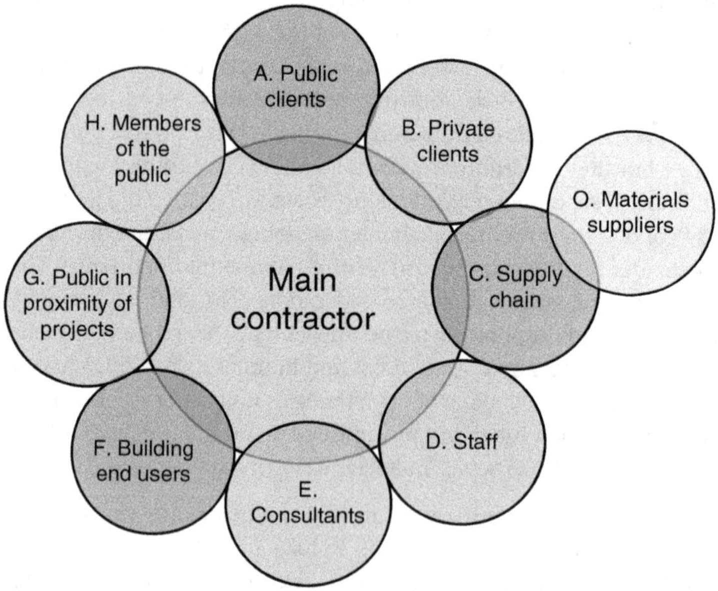

Figure 5.3 An example stakeholder map for a construction main contractor.

Different stakeholder groups may also experience different types of value. For example, a range of total value initiatives launched by a main contractor may result in some direct stakeholders achieving environmental benefits, which in turn leads to indirect stakeholders achieving social value benefits. Ethical value achieved through an initiative may also lead to well-being value achieved through the same initiative but with a different stakeholder group. Only by having a full awareness of the different types of value and the different stakeholder groups can the full positive benefits of strategies and actions be understood. Ultimately, a greater insight into the different stakeholder groups will enable a more successful rollout of a total value strategy or initiative. There will be greater clarity over the positive benefits achieved, and the different types of value experienced by different stakeholder groups will be known. Any positive benefits experienced by indirect stakeholders will also be able to be captured, and so the true value achieved can be tracked, measured and recorded.

5.5 Public Value Theory

Public Value theory is a theoretical concept that has gained significant attention within public administration, governance, and policy studies over recent years. The theory presents a framework for understanding the role of public organisations and their actions in generating value for society.

At its core, Public Value theory asserts that the primary purpose of public institutions is to create value for the public and wider society, which extends beyond financial considerations and simply a focus on the lowest cost tenders. Public Value theory is a broader concept that encompasses social, cultural and environmental factors that contribute to the overall well-being of society. It challenges the traditional public sector view that simply focusing on the lowest cost is the only method by which public organisations can evidence value for money. While it can overlap with concepts such as public goods, Public Value theory goes further by encouraging a deeper reflection on the kinds of value that a government, agency or public body can create.

Public Value theory challenges the notion that the government's role is limited to simply providing services efficiently and at a low cost. It shifts the focus from managerial goals like output maximisation to questions of what constitutes valuable outcomes for the public. The theory posits that public organisations should be guided not only by the quest for efficiency but by a broader mission of fostering democratic ideals and improving societal welfare. This can be achieved through a shift away from a focus on purely time, cost and quality, and one that encompasses a broader focus on value for society.

One of the key components of Public Value theory is its focus on the relationship between public managers and the communities they serve. This relationship is inherently political and dynamic, requiring managers to negotiate competing interests and values. Public sector management is tasked with balancing the demands of citizens, political stakeholders, and their own institutions, all while ensuring that the public's interests are protected and promoted. Public managers are therefore seen not merely as bureaucrats or administrators but as actors in a political and ethical process, responsible for creating value in ways that are consistent with the democratic framework in which they operate.

A central idea in Public Value theory is that public value should be understood in relation to citizens' expectations and preferences. Unlike traditional approaches that often emphasise efficiency or market principles, Public Value theory prioritises both public engagement and public involvement. This engagement ensures that public value is not defined solely by administrative elites but is a dynamic process informed by the diverse perspectives and needs of the population. In this sense, public value creation becomes a collaborative process, with public sector institutions working in partnership with citizens, communities and civil society to define and achieve what constitutes value.

Furthermore, Public Value theory emphasises the notion of accountability. In the public sector, accountability is critical because public institutions are funded by taxpayer money and are held to higher standards of transparency and responsibility. Public managers and policymakers are therefore tasked with making decisions that align with societal goals, ensuring that the value created is legitimate and reflective of collective interests. Public value, therefore, is not just a measure of service delivery but also a metric of the public's trust in government and its ability to meet the needs of society. The theory provides a conceptual shift in how we think about the role of government and public institutions. It highlights the importance of aligning public administration with democratic public values and citizen well-being. By emphasising the creation of value for the public, the theory advocates for a more inclusive, participatory and ethically driven approach to public governance. As such, Public Value theory remains a critical lens through which value can be assessed and refine the goals and strategies of public sector organisations.

In regard to broader value achievement and consideration, Public Value theory provides a useful framework for understanding and exploring the concept of the different types of value that can be achieved, offering insights into how public institutions and policies can contribute to the broader well-being of society. Total value extends beyond mere economic or material benefits to include intangible, cultural, environmental and relational aspects. With its focus on public benefit, Public Value theory provides a rich perspective on how public sector organisations can foster these diverse forms of value.

One of the keyways in which Public Value theory can illuminate total value is through its emphasis on the creation of value for the communities which the public body seeks to serve. Public Value theory is not simply about the efficient provision of services, but about delivering outcomes that reflect the needs, preferences, and values of the public. This notion directly connects to the idea of total value, as it places importance on the ways in which public actions contribute to societal well-being and how public services can help to promote social equity, justice, and cohesion. A focus on democratic legitimacy and responsiveness ensures that public bodies do not merely operate based on efficiency or cost effectiveness, but also on their ability to address and reflect the collective values of society, and place importance on the things that society itself values highly.

By viewing social value through the lens of Public Value theory, it can be explored how public policies and services contribute to fostering greater social cohesion and solidarity within communities. Total value achievement is inherently tied to the public's perception of how well these services serve broader social goals, such as reducing inequality, enhancing social capital and fostering trust between the public body and the intended community. For example, public policies aimed at reducing poverty, improving education or supporting identified communities can all be achieved in addition to the procurement of standard contracts (such as for the design and build of a public sector building). By expanding the procurement criteria beyond that of time, cost and quality, to include a need for the successfully awarded organisation to also deliver total value as part of their procurement requirements, the public body, through the lens of Public Value theory, can achieve more value for money overall as they will ultimately achieve a greater amount of total value for each pound spent.

Public Value theory also encourages the inclusion of multiple stakeholders in the value creation process, which is key to understanding the dynamics of total value. Public managers, who are key actors, must navigate the competing interests of different groups within society, balancing economic, social, environmental, ethical and well-being concerns (amongst others). By engaging community members in the process of defining and measuring total value, public organisations can ensure that policies are more attuned to the diverse and changing needs of the population. For instance, involving local communities in decisions about public transportation or urban planning allows for policies that reflect local priorities and a more focused total value by improving quality of life and community cohesion.

5.6 Critical Urban Theory

Critical Urban theory is a theoretical framework that is often used to both challenge and examine the structures, processes and practices that shape urban life. There is a particular focus on the issues of power, inequality and social justice in urban environments. Critical Urban theory critiques the often-dominant narratives of urbanisation and urban development and offers an alternative perspective that emphasises the role of power dynamics in shaping urban spaces and social relations. Rather than viewing cities as neutral or merely economic entities, Critical Urban theory considers them as sites of political, social and economic importance in examining the power structures, inequalities and injustices in cities.

The theory is inherently critical of urban processes that perpetuate inequalities, such as gentrification, privatisation of public space, and the displacement of low-income and minority communities. It also challenges the idea that urbanisation is a neutral or indeed an inevitable process. It argues instead that cities are products of historically specific economic, social, and political forces that serve some interests whilst marginalising others.

There are many concepts intertwined with Critical Urban theory, including the advocacy of rights for marginalised communities and the shaping of the urban environment in ways that best serve the needs of identified groups. The theory is also critical of overly privatised urban spaces and the unequal distribution of resources. It pushes for a broader approach than simply letting a capitalist market perspective determine urban environments and offers a new perspective and has become increasingly relevant in understanding the dynamics of contemporary cities, especially as urban areas around the world experience rapid gentrification, housing crises, and increased social polarisation. Critical Urban theory also provides a framework for understanding urban movements and protests, from anti-gentrification campaigns to global struggles for climate justice. It calls for an inclusive approach to urban governance, where the voices of marginalised groups are central to decision-making processes. By challenging the dominant models of urban development, Critical Urban theory contributes to a more just and equitable vision for the future of cities.

Critical Urban theory can be used to better explore the concept of achieving total value in the Built Environment. Through such a theoretical lens, the traditional structures of power, inequality, and social justice in cities can be questioned. The theory challenges dominant narratives around urbanisation, focusing on how social, economic, and political forces shape the Built Environment, and the actions of organisations can lead to creating total value for the benefit of wider society. The adoption of Critical Urban theory can help expand the idea of achieving total value and expand the idea of value itself. Through the use of Critical Urban theory, any narrow perspectives of total value can be expanded to broaden its scope in relation to urban development within the Built Environment. Since the recent resurgence of a values-driven approach, any emerging value considerations have largely been quantified and commercialised. Such an economically driven approach often leads to a narrow value focus. This focused approach can then be used to promote certain identified societal groups, although it could arguably just as easily be used to suppress one societal group at the expense of another, whilst the overall commercial 'figures' make it look like value is still being achieved. Critical Urban theory essentially allows a broader view of value – a total value perspective – to be achieved that maximises the true value for society.

Critical Urban theory is particularly focused on the ways urban processes can reinforce or exacerbate social inequality. By using this theoretical framework, it can be explored how the distribution of total value is often unequal both across cities and within cities, with certain societal groups experiencing limited access to the benefits of urban development. Through the lens of Critical Urban theory, total value can be explored as a means to address urban inequalities. This involves questioning the distribution of public goods and services – like education, healthcare, green spaces, and transportation – and how such resources are often inequitably distributed across society. A total value approach under a Critical Urban theory perspective would consider and balance one type of value that is achieved through a strategy or action against another type of value that is suppressed through the same strategy. A focus on total value considers the overall value proposition that is achieved; the use of Critical Urban theory allows this focus on total value to be broadened to consider the true impacts upon the intended societal groups.

The theory also emphasises participatory urbanism, which involves actively engaging communities in the design, planning, and decision-making processes related to the Built Environment. This participatory approach is essential in understanding and creating total value because it allows communities to have a direct say in how urban policies and projects impact their lives. This emphasis on participation aligns with the idea of total value being cocreated by communities, rather than being imposed by top-down urban policies. Through participatory approaches, communities can shape their environments in ways that foster social cohesion, resilience, and empowerment – fundamental components of total value.

5.7 Economic Value Theory

Economic Value theory focuses on how value is created, understood, and exchanged within a particular market (such as the Built Environment). This theory considers why value is placed upon certain goods and services over others, and how individual preferences influence such value.

One of the most significant contributions to Economic Value theory comes from the subjective theory of value, in that the value of goods and services is not inherent in the goods themselves but is determined by the preferences and perceptions of individuals. The value assigned to a product or service depends on its perceived utility or the satisfaction it provides. It adopts a more constructivist ontological perspective when compared to a theory such as the classical labour theory of value. This theory adopts a more objectivist ontological perspective, which I this case suggests that the value of a good is determined by the amount of labour required to produce it. The constructivist theory underscores the idea that value is not fixed or inherent but is fluid, changing based on the context of the user and their needs.

Scarcity of a resource often plays a crucial role in determining economic value. The fundamental economic problem is that the scarcity of a resource means it is unable to meet unlimited wants and needs. At a basic level, because resources are limited, choices must be made regarding their allocation. The value of any good or service is thus influenced by its availability. The scarcer a resource is, the higher its value tends to be, provided that there is demand for it. Opportunity cost is also central to value theory. This concept is of what must be given up in order to pursue one course of action over another, the use of one resource over the use of another, or in the use of time, what opportunity may be missed if time is dedicated towards a different opportunity. Every economic decision involves trade-offs, and these trade-offs reflect the value people place on their time, money or other resources.

Prices are a manifestation of value within the market. For example, the price of skilled labour within the Built Environment. Prices serve as signals of scarcity and demand. When the supply of a good is lower than demand, prices rise, signalling that the good is more valuable relative to others, and encouraging producers to supply more of it. Conversely, when supply exceeds demand, prices fall, signalling lower value or reduced consumer interest. Through this mechanism, markets allocate resources efficiently, helping to balance supply and demand. If we think, for instance, of the supply of building materials immediately preceding the opening up of societies and marketplaces post-COVID-19-induced lockdowns. Many people, fresh from the boredom of staying a home the majority of the time and staring at all the things that they wish they could improve on their properties, went straight out to buy such materials. Timber fence panels, topsoil, bricks, concrete, paint brushes, etc. all suddenly faced very high demands, but had low stock levels, and so their perceived value rose due to their scarcity.

Prices also reflect the subjective nature of value. Two individuals may value the same product differently based on their personal preferences, wealth and needs. This individual valuation is reflected in how much they are willing to pay for a good or service. In this sense, prices represent the aggregated preferences and choices of individuals in the economy, where the forces of supply and demand interact to determine the equilibrium price.

From a production perspective, value also stems from the cost and effort involved in creating a good or service. According to the classical economic view, the value of a product is determined by the amount of labour and resources used to create it. Innovation and technological advancements can also drastically alter the value of products and services. A new method of production that reduces costs or increases efficiency can lower prices, thus making a good more accessible and changing its perceived value.

In Economic Value theory, as the value of a good or service is determined by the subjective preferences of individuals, this is often at odds with the value creation-based approach that reflects the aggregation of these individual preferences across society. When considering different types of value creation, it's essential to recognise that societal preferences are diverse, and what benefits one group may not necessarily benefit another. For example, what one council needs values wise in one area (such as social value) may not be what another council needs in another area (sustainable value). Total value arises when individual preferences intersect with broader social goals. Using the economic theory of value, we can see that value is not static but is continually shaped by how individuals and groups prioritise different aspects of collective well-being.

Scarcity of a good or service will then influence the type of value that is deemed important. Such as a lower amount of educated and work-ready individuals in one location may mean that educational value achieved through social value initiatives may be highly sought after, whilst alternative locations that have high education, but low ethical behaviours may wish to focus more on enhancing their ethical value.

Opportunity cost also plays a pivotal role in assessing and understanding total value. When society makes decisions about allocating resources to one project or policy, it faces trade-offs that reflect the value of alternatives. For instance, investing in renewable energy might generate long-term environmental and health value, but it also comes with short-term costs and sacrifices, such as reduced investment in other sectors. The opportunity cost of choosing one option reflects the potential benefits that are foregone by not pursuing other valuable initiatives.

5.8 Social Constructivism

Social constructivism is a theory of learning and knowledge that emphasises the role of social interactions, cultural context, and collective experience in shaping an individual's understanding of the world. At its core, social constructivism posits that knowledge is not something that exists independently out in the world, waiting to be discovered, but rather something that is actively created through social processes. This theory suggests that what we know and understand is deeply influenced by the environment in which we live, the people we interact with, and the societal norms and practices that guide our thinking.

While social constructivism stems from several academic disciplines, it is particularly influential in the fields of education, sociology, psychology and philosophy.

Central to social constructivism is the idea that knowledge is not solely an individual achievement but is shaped by social interactions. This concept highlights that social interaction is crucial for cognitive development. Knowledge is often formed through conversation, negotiation, and shared experiences. When people interact with others, they co-create meanings, refine ideas, and develop a deeper understanding of concepts. In an educational context, for instance, students might learn not just by listening to a teacher or reading a textbook, but by engaging in discussions, debates, and group projects that allow them to challenge, reshape and build upon each other's ideas.

Another key aspect of social constructivism is the role of culture in shaping knowledge. The theory argues that what people understand as 'reality' is not universal, but instead is shaped by the cultural norms, values, language, and symbols of the society in which they live. From a social constructivist perspective, reality is something that is constructed through language and cultural practices. What one society, or individual within society, considers to be true or natural may differ greatly from another, emphasising the role of social context in constructing meaning.

In social constructivism, knowledge is not seen as a neutral or objective reflection of the world but as something actively created and shaped by social processes. This is in contrast to more traditional views of knowledge, which treat it as something to be discovered or passed on. Instead, social constructivists argue that knowledge is always embedded within social relations, language, and cultural practices.

While social constructivism has been widely influential, it is not without its critics. One concern is that social constructivism can downplay the importance of objective facts or scientific knowledge. Critics argue that by focusing too much on the social and subjective construction of knowledge, the theory risks relativism – where all interpretations of reality are seen as equally valid, potentially undermining the pursuit of objective truth. Additionally, some argue that social constructivism places too much emphasis on group-based learning and may not adequately account for individual differences in learning styles or cognitive abilities.

Utilising the theory of social constructivism to better understand total value emerges not from any intrinsic worth of a particular action or good, but through the ongoing negotiation of meaning and importance within social groups. What a society values, whether it's social justice, economic equality, environmental protection or individual and group well-being, reflects how that society constructs those concepts through social interactions.

Just as social constructivism stresses the importance of culture in shaping knowledge, it similarly highlights how cultural contexts influence what is considered socially valuable. The way different cultures define and prioritise value definition and creation can vary significantly. For example, collectivist-leaning groups may place high value on community-oriented practices, such as family cohesion, intergenerational support, and shared responsibility for well-being. These values are built through social interactions that emphasise the importance of relationships and community over individual pursuits. However, in individualistic societies, values may be more focused on personal achievement, individual rights, and self-fulfilment. The value placed on individual autonomy and freedom is socially constructed through the interactions, narratives, and norms that prioritise personal goals over group goals. Social constructivism emphasises that these values are not fixed or objective but are continuously constructed, negotiated, and sometimes contested within cultural contexts. What is considered 'valuable' or 'good' in one society might be seen differently in another, and total value, therefore, is context-dependent and fluid.

Social constructivism also emphasises the role of power in shaping knowledge and meaning. In the context of total value, this means that not all voices are equally influential in determining what is valued by society. Power structures, including political, economic and cultural hierarchies, influence how total value is constructed. For instance, dominant social groups may shape what is valued in society by controlling the narratives, norms, and symbols that define social worth. Historically, some avenues of value creation may receive more attention and focus than others due to the views and opinions of those with dominant societal views. It is therefore of paramount importance that when using social constructivism as a theoretical lens to understand total value, the decisions leading to the current value needs of society are also reflected upon to better understand the barriers that may be facing current value-focused strategies.

5.9 Total Value Theory

This chapter posits a new theory that can be utilised to explore and explain the growing need for and recognition of a total value approach – the Total Value theory. This theory describes when there is a societal or market-based shift from simply viewing value as an 'addition' to the 'core' function of a contract. For example, if a building is procured on a predominantly cost-based approach, but has 10% of its procurement criteria weighting given to social value, whilst obviously striving to make some positive difference to society or the identified demographic, it still fails to address the overall value creation of the procured organisation. A Total Value theory approach is where the entire organisation is reviewed for its overall value creation. For example, a high polluting organisation (A) that maximises an internationally based workforce with little base in the UK but that commits to donating £10,000 to a UK-based charity if successful would create less value that a predominantly UK-based organisation (B) that adopts sustainable practices but is committing to donate only £1000 to a UK-based charity if successful in the procurement exercise. In this instance, the existing sustainability focusses of organisation B, and the taxes they would pay being based in the UK, plus the economic benefit of having UK-based employees, would far outweigh the higher one-off charitable donation of organisational B.

Total Value theory is a method of full disclosure and levelling the playing field between SMEs and (multi)national organisations. The overall value created by both the organisation in its day-to-day activities and the procurement-specific value-based activities is taken into account by the procuring client. This theoretical concept can help explore and explain exactly how value is created in society and how this value is viewed by societal actors. The theory can also be used to explore short-term versus long-term value gains across society, as well as individual value-based perceptions and how such perceptions influence value creation.

5.10 Conclusion

In this chapter, the relationship between total value and a range of theoretical lenses is explored. By utilising a theoretical lens, the achievement of total value can be better understood in the context of stakeholders and how it is perceived, why it is considered legitimate (and how this legitimacy can be abused), as well as the positive contribution it can make to broader society as well as the wider economy. A new idea of Total Value theory is proposed to help better understand the modern-day interaction and expectations between organisations and society.

References

Freeman, R.E. (1984). *Strategic Management: A Stakeholder Approach*. Boston, MA: Pitman.

Suchman, M.C. (1995). Managing Legitimacy: Strategic and Institutional Approaches. *Academy of Management Review*, 20: 571-610.https://doi.org/10.5465/amr.1995.9508080331

Weick, K.E. (1995). *Sensemaking in Organizations*. Thousand Oaks, CA: Sage.

6

Technology and Value

6.1 Introduction

The following chapter spends time discussing what technological innovation means. The difference between technology hardware and software is explored, and then a range of contemporary innovations is outlined. These range from artificial intelligence (AI) to semiconductor technology. Quantum and Cloud computing to Digital Twinning (DT) to Blockchains. What the metaverse is to how blockchains work to cryptocurrency. Construction industry-specific innovations, such as Modern Methods of Construction, Platforms and Building Information Modelling, are also outlined. Each is introduced, discussed and explained, before how such technologies influence and are influenced by different types of value. This could be direct economic value, indirect social value, ethical value from decision-making or environmental value from the development process. This chapter aims to introduce the different types of technology that are and will impact the built environment, and how a value-driven approach can still be adopted when considering new technologies. Indeed, the main argument in this chapter is that value-based decision-making must be central when considering the development and implementation of new technologies.

6.2 Technology Innovation

Innovation in technology is essentially the development and implementation of new or improved technology-based solutions that ultimately enhance the way information is processed, stored, accessed, and communicated. Fundamentally, most technology innovation pertains to the way information becomes data, and how that data can then be operationalised and commercialised. Technological innovations can transform how businesses operate, how individuals interact with technology and how entire industries evolve in the digital age. At its core, technology innovation involves applying creative thinking and technological advancements to either solve existing problems or create new opportunities. It encompasses both hardware and software advancements.

Technology innovation can take the form of product innovation. This involves creating new technology products or significantly improving existing ones, with contemporary examples including smartphones, virtual reality headsets and quantum computers. Each iteration should introduce better performance, new capabilities or more user-friendly interfaces. Process innovation streamlines and optimises internal and external business processes. Automation tools, data analytics platforms and AI systems arguably improve decision-making, reduce costs and enhance productivity. Business model innovation often explores how companies leverage technology to transform the value they deliver to clients and customers. Examples of technology reshaping existing business models include cloud computing, subscription-based software and digital marketplaces. Finally, service innovation refers to technology such as online financial technology (fintech) and digital education technology (edtech). These services often improve accessibility, convenience and efficiency for users.

Value can be achieved at every stage of the innovative technology lifecycle. For example, in the research and development stage, social value can be maximised if apprentice and trainee positions are created as part of these earlier roles involved. Local value can then also be created, which can adopt the form of economic value if local supply chains are maximised for all hardware parts required. Where these cannot be sourced locally, environmental value can be created through their delivery and transportation methods, and ethical value in how these parts are sourced and the supply chain partners used. However, it is the end result where the value of any technological innovation is often realised. If we take, for example, product innovation, a more cost-effective and feature-enabled smartphone could help reduce the technology gap and digital divide in society and help bring individuals and communities online. Service innovation, such as fintech, may remove barriers to savings and investment control with members of society, allowing them to generate and build wealth, and Edtech may empower individuals to upskill and open up broader future career development opportunities. Some of the value created may be intentional, and other value created may be accidental. Some value creation may be 'obvious'. For example, developing an innovative user interface for an existing service may now mean that those with certain disabilities can access existing technology which they could not access before. Therefore, the value creation for this group is clear and intended. Yet by being aware of the different types of value and where it is created along the lifecycle of the technology innovation, firstly, additional value creation opportunities may be identified, and secondly, stakeholders can decide which products to invest in and use, based on the value created, whether this is intended value or not.

The impact of technology innovation can include benefits such as business transformation, where organisations can operate more efficiently, reach new markets and deliver better customer experiences. Job creation, for example, whilst automation may displace certain jobs, technology innovation also creates new roles and demands for digital skills and can allow existing jobs to be more efficiently completed, allowing for greater freedom, flexibility, and creativity. Global connectivity can also help economic and personal growth, as well as link to a greater quality of life for those benefiting from increased connectivity or those who now with access to technology they did not have before. Essentially, when considering value creation and enhancement due to technological innovation, it can be due to innovation in hardware or software.

6.3 Technology Hardware

Technology hardware refers to the physical components of technological systems that enable various functions, processes and services. This encompasses a wide range of devices and equipment that perform specific tasks, from basic computing to advanced scientific research. Hardware is the tangible part of technology that works alongside software to make systems operational, performing tasks such as data storage, processing, communication and interaction.

Technology hardware includes devices like computers, smartphones, servers, networking equipment, and peripheral devices, as well as specialised machinery used in industries such as manufacturing, healthcare and telecommunications. The design, production and innovation in hardware are fundamental to advancing modern technologies, influencing the efficiency, performance and capabilities of digital systems. All of which directly or indirectly underpin value creation.

The role of hardware in technology is indispensable because it provides the physical platform for software applications to function. Advances in hardware design and manufacturing influence the overall performance, speed and energy efficiency of technology systems. For instance, more powerful processors and better storage solutions enable software to run faster and more efficiently, improving user experience across devices. Hardware innovation also drives new technological developments. The evolution from early bulky computers to compact, powerful smartphones is a result of ongoing advancements in semiconductor technology, miniaturisation and energy efficiency. Additionally, breakthroughs in specialised hardware, such as quantum computing components, are opening up new frontiers for industries like medicine, finance and AI. All of these allow for a range of value creation. For instance, advancements in medicine can aid quality of life value, streamlines computer manufacturing processes can aid sustainability value by requiring lower amounts of raw materials for assembly and operation, and advancements in online capability and performance of smartphones have led to a world of training and career opportunities available to every smart phone owner increasing economic value with more people highly trained and able to generate multiple income streams or be able to have a greater work-life balance.

The use of hardware has also enabled greater data collection when it comes to value measurement and benchmarking. Employees with responsibility for value delivery at organisations now have access to tablets and phones for data collection. For example, recipients of value-creating initiatives are now able to be surveyed in real time for feedback. Feedback that can then be critically assessed and reviewed, with changes implemented for the further creation of value.

6.4 Technology Software

Technology software refers to the programs, applications and systems that run on hardware to perform a variety of functions. Unlike hardware, which is the physical component of a computer or electronic device, software is intangible and consists of the code and instructions that tell the hardware how to perform specific tasks. Software is integral to virtually all technological systems, from basic applications to complex, industry-specific solutions, and it plays a central role in shaping the functionality, efficiency and experience of users interacting with technology.

System software serves as the foundation for running application software and managing the computer's hardware. It allows the user to interact with the computer and ensures that various hardware components work in harmony. Operating Systems are perhaps the most critical system software. The Operating System acts as an intermediary between the user and the hardware, managing resources such as memory, processing power and input/output devices. Device Drivers are specialised programs that allow the operating system to communicate with specific hardware components like printers, keyboards, and graphics cards. Without drivers, devices would not function properly. Utility Software includes programs that manage, maintain and optimise the system's functionality, such as antivirus software, disk cleanup tools and backup utilities.

Application software is designed to perform specific tasks or solve particular problems for users. It utilises the capabilities of the system software to carry out its intended functions. Examples of application software span across various fields and industries and can include productivity software. Programs like Microsoft Office and Google Workspace help individuals and businesses create, edit and manage documents, spreadsheets, presentations and more. Creative software includes software designed for creative professionals, and these applications are often used in graphic design, video production and other media-related fields. Web browsers allow users to browse the internet, access websites and use online services. Navigating online in a way that is efficient and effective. Enterprise software allows business solutions to manage operations, finances, and customer relationships. However, all software can essentially be considered as a 'vehicle' by which users can access or organise information in an efficient and effective manner. Therefore, the use of software can itself create and enhance value. This may result in some technologically innovative companies creating value, not through any additional activity, but through the very business operations that embed and utilise. This is fine. It should be encouraged. Too often, 'value' has been viewed as something organisations bring that is in 'addition' to their day-to-day operations, but the core business operations should be considered with just as much focus and emphasis with regard to additional value creation.

Technology software plays a crucial role in the functionality of modern devices and systems. The interaction between software and hardware determines the capabilities and performance of computers, smartphones and other digital devices. Without software, hardware would be unable to perform any meaningful tasks. Software allows for the automation of processes, reducing human error and improving efficiency. For example, business management software streamlines accounting, inventory tracking, and human resources tasks. Software also enables users to tailor devices and systems to meet their specific needs. This could range from personalising a smartphone to selecting specialised software tools for a professional field. Finally, software is the driving force behind technological innovation. It enables advancements in fields like AI, machine learning (ML), robotics and virtual reality, transforming industries and everyday life, leading to continuous opportunities for value creation and enhancement.

6.5 Artificial Intelligence and Machine Learning

AI and ML are two of the most transformative technologies in the world today. From voice assistants and recommendation systems to self-driving cars and healthcare diagnostics, AI and ML are reshaping industries and daily life. Though often used interchangeably, they refer to related but distinct concepts. AI is the broader field focused on creating systems that can mimic or exceed human intelligence, while ML is a subfield of AI focused on teaching machines to learn from data without being explicitly programmed.

AI refers to the ability of a machine or computer system to perform tasks that typically require human intelligence. These tasks include reasoning, learning, problem-solving, perception, language understanding and decision-making. AI aims to simulate human cognitive functions and automate intelligent behaviour. AI can be classed as narrow AI, which is AI that is trained and designed for a specific task, such as image recognition, translation or playing chess. Most AI in use today is narrow AI. There is also general AI, which refers to machines that can perform any intellectual task a human can do, with the ability to reason, understand context and learn from diverse experiences. This level of AI remains largely theoretical at present. Indeed, even narrow AI is fairly limited in its current abilities. Despite the claimed benefits and competitive advantages that numerous companies are claiming AI brings to them at present, it is essentially a very efficient data management and structuring tool. AI systems often use a combination of rule-based algorithms, logic programming, natural language processing, robotics and, increasingly, ML to function effectively.

ML is a subset of AI focused on developing algorithms that enable computers to learn from and make predictions or decisions based on data. Instead of being explicitly programmed for every task, ML systems identify patterns in data and use these insights to improve their performance over time. ML is based on the idea that systems can automatically learn and adapt without following strict instructions. This makes it extremely useful for complex tasks where traditional programming is impractical.

There are three main types of ML (Table 6.1):

Table 6.1 Types of machine learning.

Supervised learning	In supervised learning, the model is trained on a labelled dataset, which means each input is paired with the correct output. The algorithm learns to map inputs to outputs and is evaluated based on its accuracy in predicting outcomes, e.g. email spam detection, house price prediction and image classification.
Unsupervised learning	Unsupervised learning involves training models on data that has not been labelled. The goal is to identify underlying structures or patterns in the data. E.g. customer segmentation, anomaly detection and topic modelling in text analysis.
Reinforcement learning	This type of learning is inspired by behavioural psychology. An agent interacts with an environment, learning to make decisions by receiving rewards or penalties. Over time, it learns the optimal policy for maximising its cumulative reward. E.g. game-playing agents, robotics and autonomous driving systems.

Deep learning is a specialised form of ML that uses neural networks with multiple layers, hence 'deep' learning. These deep neural networks are inspired by the structure of the human brain and are particularly effective at tasks like image recognition, natural language processing, and speech recognition. Deep learning models require large amounts of data and powerful hardware but have achieved state-of-the-art performance in many domains. This includes industries such as healthcare, where AI models assist in diagnosing diseases from medical imaging. ML also helps in predicting patient outcomes and personalising treatments. Chatbots and virtual nurses also help improve patient engagement and support – or at least make some elements of patient interaction more efficient, even if a somewhat pessimistic view is that they don't enhance engagement or patient satisfaction.

In the finance industry, ML can aid in fraud detection systems to identify unusual transaction patterns. AI can also help with credit scoring, algorithmic trading, and customer service automation. Risk modelling and financial forecasting also benefit from ML algorithms through pattern identification and automated simulation modelling of outcomes and scenarios.

Retail and E-commerce use ML to personalise user experiences, AI-powered chatbots are again used but to handle customer queries in real time, and inventory management and demand forecasting are optimised through predictive analytics. Transportation and logistics also use ML in instances such as self-driving cars that rely on AI and ML to understand surroundings and make decisions, route optimisation and demand prediction improve efficiency in logistics, with modelling and simulations employed to enable the most efficient routes to be utilised, and drones and robots assisting in last-mile delivery. Education also utilises personalised learning platforms to adapt to individual student needs using ML. AI tutors and grading systems save teachers' time and provide real-time feedback, and predictive analytics can identify students at risk of dropping out.

All of these initiatives and uses of AI and ML provide ample opportunities for value creation and enhancement. The initiatives themselves, such as personalised feedback delivered to learners in a more efficient manner with regard to teacher time, and logistics companies not spending as much time in traffic and so reducing their CO_2 generation – all generate or enhance value. The increasing prevalence of AI and ML is intrinsically linked to the greater creation of value. The more such software tools are used, the more potential there is for value creation and enhancement.

Despite many potential benefits of AI and ML for broad value creation and enhancement, there are still many challenges and ethical implications that must be considered. This includes bias and fairness as AI systems can inherit biases from the data they're trained on. If not addressed, this can lead to unfair treatment, for example, if AI is used in areas like recruitment. The software utilised could draw upon unconscious bias patterns in the data to continue to make errors and decisions that humans which could lead to the exclusion of certain groups, thereby undoing any potential value-based benefits. Data privacy is also a concern as many AI applications require vast amounts of data, raising concerns about user privacy and data security. Again, this could undermine any value creation exercises if data is breached or not secured in an appropriate manner. Job displacement is a potentially huge area that will need to be considered through the lens of total value creation. AI and general automation open up the potential for jobs to be reduced, removed, and moved abroad. Such actions will then need to be reconciled against more local value creation. For example, the lack of local UK-based recruitment may be offset by international-based recruitment, but for a local social value agenda, this will lead to international value generation instead of local value generation.

6.6 Semiconductors

Semiconductors are materials that have electrical properties somewhere between conductors (like copper) and insulators (like glass). They form the foundation of modern electronics, powering everything from smartphones and laptops to cars and satellites. At the heart of this technology are tiny components like transistors, diodes and integrated circuits, all of which rely on the unique behaviour of semiconductors to control the flow of electricity. The most commonly used semiconductor material is silicon, which is abundant, stable and easy to manufacture at scale. Other materials, such as gallium arsenide, germanium and more recently, compound semiconductors and materials like graphene, are used for specific applications that demand higher speed or performance.

What makes semiconductors special is their ability to switch between conducting and insulating states. This is achieved by introducing impurities into the pure semiconductor material in a process called doping. Doping adds free charge carriers, either electrons (negative) or holes (positive), that change how the material conducts electricity. A p-type semiconductor has more positive charge carriers, while an n-type semiconductor has more negative charge carriers. When these two types are combined, they create a p-n junction, which is essential for building diodes, transistors and other basic electronic devices. The most important semiconductor device is the transistor, which acts as a tiny electronic switch. Billions of these transistors are packed into modern computer chips, allowing them to perform calculations, store data, and run complex software. Advances in semiconductor technology, such as the development of smaller and faster transistors, have fuelled the exponential growth of computing power. Semiconductors are also critical to energy-efficient technologies. Power semiconductors regulate energy flow in everything from electric vehicles to solar inverters, reducing energy loss and improving performance. Optoelectronic semiconductors enable light-based technologies like LEDs, laser diodes and image sensors found in cameras and medical devices.

The semiconductor industry is a global ecosystem, involving design, fabrication and packaging. Fabricating chips requires highly advanced facilities called fabs, where wafers of semiconductor material are processed using photolithography, etching, ion implantation and other precise techniques to build circuits at the nanometre scale. In recent years, the demand for semiconductors has surged due to the rise of AI, 5G, electric vehicles and the Internet of Things (IoT). This growth has also highlighted geopolitical issues, as many countries seek to secure domestic chip manufacturing capabilities to reduce reliance on foreign supply chains. The potential of semiconductors for total value creation and enhancement is significant. This technology spans numerous value types and has the power to both create value through its use in technology aimed at value creation and create value through the development and procurement of semiconductors themselves.

6.7 Quantum Computing

Quantum computing is a revolutionary type of computing that uses the principles of quantum mechanics (the physics that governs the subatomic world) to process information in fundamentally new ways. While classical computers rely on bits, which can be either a 0 or a 1, quantum computers use *quantum bits* or *qubits*, which can exist as 0, 1 or both at the same time due to a property known as superposition. Superposition allows a qubit to perform multiple calculations at once. For example, while a classical bit processes one possible input at a time, a qubit can explore many possible inputs simultaneously. This massively parallel processing power gives quantum computers the potential to solve certain complex problems much faster than even the most powerful supercomputers today.

Another key quantum property is entanglement, which enables qubits to become linked so that the state of one qubit instantly influences the state of another, no matter how far apart they are. Entanglement allows quantum computers to handle and store information in highly correlated and powerful ways that classical systems simply cannot replicate. Quantum computers perform operations using quantum gates, which manipulate qubits through controlled changes in their states. These gates form quantum circuits, which are the building blocks of quantum algorithms, specially designed procedures that take advantage of quantum phenomena to achieve results faster than classical algorithms. Although quantum computing is still in its early stages, there are several promising uses.

The potential use of quantum computing in overcoming many problems faced today is of great potential. For example, problems such as route planning, financial modelling and supply chain management can all benefit from quantum computing. Quantum algorithms can explore many possible configurations simultaneously, potentially finding optimal solutions far faster than traditional methods. Despite its promise, quantum computing faces major challenges. Qubits are extremely delicate and can be easily disturbed by their environment, a problem known as quantum decoherence. To address this, researchers are developing quantum error correction techniques and building quantum computers in highly controlled environments, often at temperatures close to absolute zero. Several types of quantum hardware are being explored, including superconducting qubits, trapped ions and photonic systems. Each approach has advantages and trade-offs in terms of scalability, stability, and speed. However, each approach still retains the potential to create and enhance a range of values. Whilst the increased energy required to research and develop such innovations may provide short-term sustainability issues (in the use of more carbon), the long-term sustainability benefits of more efficient technology use will (hopefully) offset any short-term negatives. A greater focus on technological innovations may also lead to more long-term and highly paid secure job prospects, which would create both social value in terms of job creation and economic value in terms of retained and enhanced wealth for organisations.

6.8 Cloud Computing

Cloud computing is the delivery of computing services over the internet, enabling users to access and use technology resources, such as servers, storage, databases, software and networking, often on a pay-as-you-go basis, without having to own or maintain the physical infrastructure. Instead of relying on local servers or personal devices, cloud computing allows individuals and organisations to store, manage and process data remotely, offering scalable and flexible solutions that can be accessed from anywhere with an internet connection.

Cloud computing has revolutionised how businesses and individuals manage and use technology by eliminating the need for on-premises hardware, reducing costs, and providing easier access to advanced software applications and data storage. The main features of cloud computing include allowing users to manage resources like storage, computing power and networking services automatically without requiring human intervention from the cloud provider. This flexibility gives users the ability to scale up or down based on their needs in real time. Cloud services are also accessible over the internet using a range of devices such as computers, smartphones and tablets. Users can access data and applications from anywhere, making it convenient for remote work and collaboration. Cloud providers also use multi-tenant models, where resources such as servers and storage are pooled together and shared amongst multiple customers. This sharing ensures efficiency and scalability while reducing the overall cost of infrastructure. Cloud computing resources can also be quickly scaled up or down depending on demand. This elasticity allows organisations to handle sudden increases in traffic or workload without worrying about running out of resources or overspending on infrastructure.

The benefits of cloud computing are vast and include cost efficiency, as cloud computing eliminates the need for businesses to invest in expensive hardware, reducing upfront costs and ongoing maintenance expenses. The 'pay-as-you-go' pricing model makes it more affordable for businesses of all sizes. Scalability and flexibility are other benefits, as cloud computing services can easily scale to meet changing demands. Businesses can expand their resources during peak times and reduce them during slower periods without overcommitting resources. This is essential in the 'subcontracting' focused built environment where organisations essentially embrace the idea of 'lean' operating, and only retain the core management staff required, and then adopt a subcontracting model on a project-by-project basis when required. Somewhat safe in the knowledge that without ongoing active projects, the 'core' staff base only extends to the few directly employed management staff. This same principle can be adopted in the use of technology (both hardware and software), with licences available that only need to be paid for when used. Whilst once cloud computing is used, it would perhaps be inefficient to then cancel subscriptions entirely and lose the data stored, it would be prudent to only pay for the minimum amount of storage capacity needed. Accessibility and collaboration are further benefits of cloud computing, as the software allows teams to access data and applications from anywhere, improving collaboration and productivity. This accessibility is particularly beneficial for remote working with increasingly international and multidisciplinary built environment teams. Finally, the security and backup capabilities of cloud computing are of increasing benefit in a somewhat increasingly uncertain economic and geopolitical climate. Cloud providers offer advanced security measures like encryption, authentication and regular data backups to protect user data. Cloud services also provide disaster recovery capabilities, reducing the risk of data loss.

However, security and backup processes of data can equally be a downside to cloud computing in this same current climate. Security and privacy have always been considered of paramount importance across all industries. With commercially sensitive data often the difference between procurement success and failure, secure storage is of organisational critical importance. Storing sensitive data in the cloud raises concerns about data privacy and security breaches. It's important for users to select reliable cloud providers with robust security protocols. However, unfortunately, even then, high-profile data hacks, leaks and deletions can occur. Reliability issues are also a disadvantage. Although cloud services are generally reliable, downtime or service interruptions can occur. Built Environment organisations need to assess the reliability of their cloud provider and have contingency plans in place. Data transfer and bandwidth costs also need to be considered and could be prohibitive in nature. Moving large amounts of data to and from the cloud can incur additional costs and time delays, especially for businesses with high data transfer needs (such as complex drawings, extensive scope documents and any modelling undertaken).

Nevertheless, cloud computing aligns with creating value in the Built Environment. Notwithstanding the life cycle data storage costs and carbon involved.

6.9 Digital Twinning

DT is the virtual creation of a physical object, system or process. In its true sense and potential, DT uses real-time data and simulations to mirror a real-world counterpart, enabling monitoring, analysis and optimisation. Such simulations can extend to modelling past and future scenarios. This is of paramount importance to the Built Environment. Firstly, the modelling of past scenarios can help understand disaster events and reveal how built assets performed under unusual and previously unexpected circumstances. Such lessons can then serve to inform the development of future Built Environment assets. This can help maximise leadership and ethical value, as well as the value placed on health and safety and well-being. For example, future building occupants can be safe in the knowledge that their building has been informed in its design and construction by lessons learnt from thousands of past environmental events and previous building successes and failures. Future scenarios can also be modelled and simulated to reach the same outcomes, in a building designed that withstand all simulated events thrown at it.

DT is increasingly common in manufacturing, healthcare and smart cities, and digital twins themselves can help improve performance, predict failures and support better decision-making through continuous feedback and insights. DT is a transformative technology. The digital model mirrors the real-world counterpart in real time, using data from sensors and other sources to simulate its behaviour, condition and performance. The concept of DT has evolved rapidly with the advancement of other technologies discussed, such as the IoT, AI and cloud computing. Originally developed for aerospace and manufacturing, DT is now being applied across many industries, including healthcare, energy, automotive, urban planning, and the Built Environment. The core idea is to link the physical and digital worlds, enabling real-time monitoring, testing, analysis and optimisation without physically intervening in the real object or environment.

A digital twin works by integrating data from a physical object or system with a virtual model. This data typically comes from sensors embedded in the physical entity, which collects information on a variety of factors through a variety of data collection points, such as temperature, motion, pressure, energy use, wear and other variables. The collected data is then fed into software platforms that update the virtual model in real time. For example, a digital twin of a printer can track the usage, performance, and maintenance needs of the real-world printer. It can schedule updates when not in use, dispatch ink cartridges when needed, and arrange for proactive maintenance to prevent unscheduled downtime.

Engineers can use the digital twin to predict failures before they happen, simulate different operating conditions or test design modifications, all without interrupting operations. In the Built Environment, this example could be extended to a whole building. The digital twin can simulate upcoming forecast weather events, the impact that increased footfall will have on carpet tiles, and the accurate lifespan of glazing based on current conditions. Mechanical and heat recovery systems could also be analysed and changes in their settings tested in the digital environment, with the most efficient course of action then automatically happening in the real-world building, which is linked to and directly controlled by the digital twin.

The key components of DT include the:

- Physical Object or Process: the real-world asset being replicated, such as a vehicle, factory or bridge.
- Digital Model: a computerised version of the physical asset that includes its structure, dynamics and operational behaviour.
- Data Integration: real-time data from sensors and devices that continuously update the digital model to reflect the current state of the physical asset.
- Analytical Tools: algorithms and ML models that interpret the data, predict future outcomes and offer actionable insights.

The application of DT technology and processes is vast. It includes sectors such as manufacturing, where factories can use digital twins to optimise operations, detect faults early and improve productivity, and entire production lines can be modelled and subjected to simulations to identify bottlenecks and test new workflows. In healthcare, DT technology of patients can help simulate medical procedures, personalise treatments and monitor chronic conditions remotely. In the energy sector, power plants and grids can use digital twins to optimise energy flow, monitor wear on components and reduce downtime through predictive maintenance. In the automotive and aerospace sectors, engineers build digital twins of vehicles and aircraft to test performance and enhance safety and design next-generation products more efficiently.

The benefits of DT are equally vast. Predictive maintenance can be undertaken by the continuous monitoring of the real-life asset using appropriate sensors and equipment, allowing the digital twins to forecast failures before they occur, reducing downtime and repair costs. Improved decision-making with real-time data based on simulations that then enable better insights, better risk assessments, and greater planning. Benefits also include cost savings that can be gained from the virtual testing, which then reduces the need for physical prototypes and expensive trial-and-error experiments. There is also the benefit of enhanced innovation in that designers and engineers can quickly test new ideas in the digital environment, speeding up development cycles.

However, the increasing use and ultimate dependency upon digital twins come with challenges and considerations. For example, implementing and maintaining accurate digital twins requires significant investment in sensors, connectivity and data infrastructure. The trust, accuracy and dependency of the data are also a concern. Inaccurate data, even if only slightly, can throw off the reliability of simulations. The cybersecurity of the data and model itself is another concern, as these systems often deal with sensitive operational data, especially if the complete data set and plans of a physical asset are all stored in one place.

If access to such data fell into the wrong hands, this could be a major security threat and undermine trust in the entire DT process and technology.

6.10 Metaverse

The metaverse is essentially a series of connected digital twins of physical assets. These digital twins could be held in a virtual world in which users navigate their way from one physical asset to another. The metaverse structures and presents all digital twin models, and could be on a full city scale. Therefore, the metaverse shares many of the advantages and challenges of single digital twins, but on a larger scale. However, one problem that users may encounter in the future with the metaverse is if more than one was to exist. For example, similar to television subscription services, there may be multiple metaverses, each housing virtual replicas of different physical assets. In theory, such metaverses could present whole new markets in value. For example, digital replicas of significant buildings, when combined with technology such as virtual reality headsets, can allow those unable to travel to such significant buildings to still experience them, but in a digital world. Equally, historical buildings may be preserved, allowing people to visit in future. Such value can be not only in the preservation of the building itself but also in the educational opportunities provided to individuals.

6.11 Blockchains

Blockchain is a decentralised, digital ledger technology that records transactions across multiple computers in a way that ensures security, transparency and immutability. Originally developed as the underlying technology for Bitcoin, blockchain has since evolved far beyond cryptocurrency and is now used across various industries for secure and transparent data management. At its core, a blockchain is a chain of blocks, with each block containing a set of transactions or data. Once a block is filled with data, it is 'chained' to the previous block using cryptographic methods, forming a continuous and unchangeable ledger. This chain of information is distributed across a network of computers (called nodes), each maintaining a copy of the entire blockchain. This decentralised structure makes blockchain highly secure and resistant to tampering.

The key characteristics of blockchain include its decentralised nature. Unlike traditional databases managed by a central authority, blockchain operates on a peer-to-peer network. Every participant (node) in the network has access to the entire ledger, and changes to the data must be validated through a consensus mechanism. Blockchains also afford transparency in that all transactions on a public blockchain are visible to every participant. This transparency builds trust amongst users, especially in systems where verification is crucial. Immutability of data results from once data is recorded in a blockchain and confirmed by the network, it cannot be changed or deleted. This makes the technology ideal for maintaining accurate and tamper-proof records. The security of blockchains is also a benefit as they use cryptographic techniques to secure data. Each block contains a unique code called a hash and the hash of the previous block, linking them together. Any attempt to

Table 6.2 Blockchain stages.

1	A user initiates a transaction.
2	The transaction is broadcast to a network of nodes.
3	The network validates the transaction using a consensus algorithm, such as Proof of Work (PoW) or Proof of Stake (PoS).
4	Once validated, the transaction is bundled with others into a block.
5	The block is added to the chain and distributed across the network.
6	The transaction is complete and permanently recorded.

alter the data changes the hash, which is immediately recognised by the network as invalid. Essentially, a blockchain works through the high-level stages (Table 6.2):

Blockchains can be applied across many sectors, including arguably the most well-known of cryptocurrencies, to record and verify transactions. However, blockchains can also be used across supply chain management to provide end-to-end visibility, helping track the origin, condition and location of goods. Healthcare to secure the sharing of medical records between providers while maintaining patient privacy. Finance to in areas such as smart contracts to automate and enforce agreements, reducing the need for intermediaries, and also in voting systems to have transparent and tamper-proof voting platforms to ensure election integrity. Whilst still in mainstream infancy, the potential of blockchains to create value is huge. The transparency and accuracy of the technology are huge benefits to trust value being achieved almost as standard when interacting via blockchain technology.

6.12 Cryptocurrency

Closely aligned to the technology of blockchains is that of cryptocurrency. Cryptocurrency is a form of digital or virtual currency that uses cryptography for security. Unlike traditional currencies issued and regulated by governments. Cryptocurrencies are decentralised and typically operate on blockchains. The concept of cryptocurrency was first introduced in 2009 with the launch of Bitcoin, which was designed as a peer-to-peer electronic cash system, allowing people to send and receive payments without the need for banks or financial intermediaries. Since then, thousands of other cryptocurrencies have emerged.

Cryptocurrencies are stored in digital wallets, which can be software-based (online or mobile apps) or hardware-based (physical devices). These wallets allow users to send, receive, and store their digital assets securely using private keys.

The benefits of cryptocurrency include its decentralised nature, in that it eliminates the need for intermediaries like banks, enabling faster and cheaper transactions. It also offers security and transparency in that blockchain technology ensures secure and visible records of all transactions. One of the main benefits that relates to achieving value is the financial inclusion of cryptocurrency. Such inclusion provides access to financial services for unbanked and underbanked populations, thus potentially providing economic value to large parts of the global population, helping lift them out of poverty by providing access to a means to earn, save and spend money securely.

However, many high-profile risks and challenges do exist. For example, the well-known volatility of cryptocurrency markets can lead to high fluctuations in the value of currency held. There is also much regulatory uncertainty around cryptocurrency as the legal status varies across countries, with ongoing debates about regulation and ethical use and 'mining' of such currencies. There are also large-scale and ongoing security threats. Users have often reported missing cryptocurrency or being targeted in real life and forced to hand over access to their crypto banking accounts and find their entire stock stolen from them. There is also an increasing awareness of the intense environmental impact involved with cryptocurrencies. They consume large amounts of energy, which often then undermines any value benefits achieved due to the high environmental costs.

6.13 Conclusion

This chapter discusses just a few of the technologies that are currently impacting the Built Environment and will impact the built environment in future. Despite the varied use and application of the technologies, they all have many similarities. Most notably, they all have the potential for industry-changing Total Value creation. If developed and utilised correctly, each technology described can create efficient, effective and revolutionary value on a local and global scale. However, all the technologies come with both immediate and longer-term risks. Risks that may undermine any value achieved in other areas. The technologies are also often interrelated. Advancements in hardware, for example, are needed to maximise the use of software. The rise in cryptocurrency is closely associated with the levels of trust in blockchain technology. Metaverse technology needs accurate digital twins in order to be successful, whilst the digital twins themselves depend on the accurate data, which in turn is linked back to the availability of the hardware components to collect such data. The data collection and transmission are then highly dependent upon cybersecurity being in place to protect the data generated. The rise in technology, therefore, does offer potential for new value to be created and existing value to be enhanced. However, it cannot help but be noticed that a failure in one area of technology development could undermine the value created by all other technology developments.

Part II

Value Action

7

Measuring Value

7.1 Introduction to Chapter

This chapter of the book will start by articulating what social value measurement comprises and why it is important for different stakeholders. It will clarify and distinguish between project outputs and outcomes and provide examples to explain the differences. It will outline why social value creation is particularly related to outcomes and the importance of quantifying these to gauge success. The chapter will then outline why measurement of social value is critical for individuals, teams and organisations in their decision-making and the benefits that it can have. Thereafter, it will identify and discuss the various current approaches to measuring social value, including cost-benefit analysis, social accounting, social return on investment and basic efficiency resource analysis. For each of these approaches, the chapter will articulate the calculation methodology using examples where necessary. Finally, it will discuss the 'TOM system' as an alternative way to measure social value and the different approaches this has taken to traditional methodologies.

7.2 What Is Social Value Measurement?

Social value measurement attempts to comprehend and record the relative importance that individuals, groups or organisations place on changes relating to well-being that they experience. In this way, it can assist in better informing and influencing the decisions that are made to promote positive effects whilst reducing negative effects. Measuring social value is important as businesses can then have a good estimation of the impacts that are brought about through their interventions and decision-making. These impacts may be positive or negative but understanding them can help better assess the relative risks and opportunities that they can create. Social value measurement relies on involving key individuals who may be stakeholders from different areas of the business. They should be listened to and, wherever possible, take account of their views and suggestions.

7.3 Measuring Social Value: Current Measurement Techniques and Practices

For many years, it has become increasingly important to attempt to quantify the social value and social impact made by organisations through measurement. The reasons for wanting to quantify value in this respect emanate from many different angles. Investors and funders, for instance, will want to be sure that their money is being channelled into the projects which are calculated as being the most effective. In a similar way, governments, policy makers and other public sector bodies are accountable for their decisions around spending and must demonstrate value for money through regular audits. In addition, those organisations which perform social functions, such as charities, also have responsibilities to ensure that funds are expended on projects that have beneficial outcomes and tangible real impacts. This has led such organisations to review their traditional business models away from purely financial accounting into social value assessment. This has had the effect of generating more interest in concepts around 'value added' and 'value for money' as strategic objectives and key criteria for measuring their performance against. In assessing performance in this way, it is important for organisations to distinguish between outputs and outcomes. Outputs relate to the organisational activities, normally as part of planned objectives or action plans. Examples in a social sense may include providing a young person's community centre as part of a new development. Conversely, outcomes are longer-term and look at the overall social effects that the outcomes have had over time. For example, using the same example, providing the young person's community centre may have reduced young people on the street and had an influence on reducing adolescent criminality, such as vandalism in the local area. If the community centre offers a facility for career advice, then the longer-term outcome could be to improve young people's ability to pursue careers and thus reduce youth unemployment in the local area.

Social value is concerned with those outcomes that potentially change people's lives and improve their well-being, rather than just outputs linked to initiatives that may have no longer-lasting effects on individuals or communities. Social value assessment revolves around the measurement of the wider outcomes that are directly attributable to the actions of an organisation. These assessments need to take into account what may have happened anyway through the actions of other organisations and stakeholders, e.g. local authority. Demonstrating social value in this way can be hugely important for organisations, especially in a period of recession where there could be government spending cuts and where competition for finite resources is high. However, it is sometimes difficult to measure and quantify the effects of these outcomes, owing to their non-financial nature. For this reason, social value outcomes are sometimes referred to as 'soft outcomes' owing to the challenges faced with quantifying and measuring them. Accordingly, organisations should be aware of the time and money that they may be required to invest to undertake such assessments.

7.4 Benefits of Measuring Social Value

Measuring social value allows organisations to justify spending and their investments in a particular area by looking at the impact this has had. In this way, it gives them the ability to defend their roles and decision-making in providing the services they have undertaken. In doing so, it gives them the ability to generate more funding by demonstrating that previous funding into a particular area has made a difference. Furthermore, it allows organisations to closely focus their resources and efforts on a specific area as a targeted rather than a scatter gun approach. This facilitates a more controlled approach where they can better plan and allocate the resources they need to invest in a project or initiative. In addition, a systematic approach of constantly reviewing performance in this way can allow organisations to continually improve by asking themselves the question of what could be done differently to generate more impact in future. Having metrics which can demonstrate successful impacts through social value assessment can assist in communication strategies in being able to convey to employees, investors, communities and other stakeholders what has been achieved.

7.5 Current Approaches to Measuring Social Value

Sometimes it is difficult for organisations to agree on what social value is, owing to its subjective nature and the potential for it to mean different things to different people. Saxon (2005) advocated that different models of value emerge depending on who the stakeholders are. For instance, on a commercial development, a developer will assess the project's worth through calculating its estimated cost and comparing this to the asset value they hope to achieve, coupled with the anticipated risk involved. Local authorities, especially town planners, will look at the value of the same development through the lens of the community interests in the local area, including the social benefits, concerns and objections of local people. Conversely, investors and funders will normally look at return on their investment and any inherent risks associated with recouping their money. Finally, owner occupiers may view value more from an in-use perspective over possibly a longer period of time. Owing to this dilemma, the prospect of measuring it presents a particular challenge. Although there are many different ways and approaches to measure social value, this book will concentrate on four methods below.

7.5.1 Cost-benefit Analysis

Cost-benefit analysis revolves around decision-making to determine whether a particular project is worthy of investment, whilst providing a basis for comparing projects. This involves looking at the expected projected benefits against the costs associated with their implementation. If the projected benefits outweigh the costs, then this provides a financial justification for implementing a project. Normally, this economic-based analysis would look at a range of different project alternatives, by way of a business case, including an option to 'do nothing'. This method heavily relies on collating data associated with both the benefits and the costs. Costs may include new buildings, staff wages, equipment, servicing and interest on borrowing. Projected benefits may include advantageous effects in the local area and on the communities that live in it. Another outcome could be job creation and positive effects on the well-being and lifestyle of individuals, including job prospects and revenue.

7.5.2 Social Accounting

Social accounting techniques are commonly used in the corporate business sector to measure corporate social responsibility. It is a technique that is also commonly used by charities and foundations to measure outcomes. The main difference between this method and cost-benefit analysis is that it applies to the organisation as a whole rather than a particular project. It utilised all the existing resources within the business and existing reporting and monitoring systems to incorporate social accounting within. In this way, it fully integrates social accounting into its primary business activities and existing governance structures. Social accounting will involve providing clarity on what organisations are attempting to achieve and how they intend to undertake this. Thereafter, the measurement process involves generating both quantitative (numeric) and qualitative (descriptive) data relating to the organisation's aims and objectives. This data is reported within social accounts, which are independently audited on an annual basis. For this reason, organisations are assessed on more than just their finances and as such, this broadens the scope of traditional accounting practices. There are broadly four steps to implementing social accounting, and these are contained in Figure 7.1.

Step One will include carefully setting out what values, vision and objectives related to what difference the organisation wishes to make are absolutely clear and understood. Step Two will involve putting a plan in place and obtaining data on project outcomes and key performance indicators (KPIs), with consultation with all stakeholders. Step Three will involve preparing social accounts through data gathered in Step Two and reporting on outcomes against benchmark targets. Finally, Step Four will involve having draft accounts completed and signed off by a social auditor.

7.5.3 Social Return on Investment

The social return on investment method of measuring social value was generated from both the cost-benefit analysis and social accounting methodologies. It differs from social accounting in that it places a financial value on the wider impact that organisations make on social value. Sometimes this financial impact has been referred to as having a 'multiplier effect'. The regeneration of East Manchester in the UK, between 1999 and 2009, which Professor Challender was involved in, utilised this principle. This neglected and heavily dilapidated area of Manchester received both public and private sector funding to create a new college campus, football stadium, cycling arena, shopping area and library over a

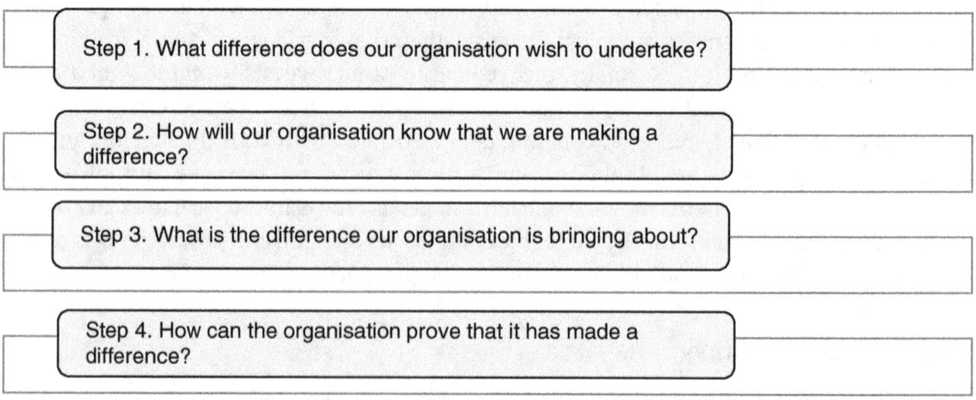

Figure 7.1 Four main steps to social accounting.

10-year period. It was calculated that for every £1 spent on this development, the overall wider outcome could generate £2.5 based on a multiplier of 2.5. This multiplier was arrived at considering the new service sector that would be developed alongside the jobs which would be created and the wider regenerative benefits for the local community. The creation of parks, public realm and GP surgeries were all linked to well-being and health, and these benefits also went into the calculation for the multiplier. This method of transporting wider benefits into financial outcomes proved particularly beneficial in justifying and obtaining funding from both public and private sectors.

7.5.4 Basic Efficiency Resource Analysis

This method of assessment represents a more recent cost-effective approach to social value and is one that has been used by the UK charity Oxfam in the past. This assessment method is intended to generate a concise framework for evaluating activities, campaigns or programmes which are complicated and challenging. It measures the impact compared to the resources required to generate the impact. In addition, it utilises a process whereby there is a relative assessment is made on performance when compared to other available initiatives. This process works by quantifying each initiative as a unit in its own right and comparing the unit against other units. Units in this sense may comprise work areas or a team. Inputs and outputs are closely recorded, and these may be facilitated through group discussions and interviews with stakeholders. The basic efficiency resource analysis is a model that does not simply measure value or impact but looks at how efficiencies can be generated. This makes it a particularly proactive method of assessment to help organisations improve their ongoing performance levels. They can be comprised of many different facets, including rituals, traditions, customs and beliefs.

7.6 Alternative Ways of Measuring Social Value

The social value 'TOM System' has been introduced by one of the leading consultants on social value in the UK, namely the Social Value Portal. The TOM system encapsulates Themes, Outcomes and Measures into its methodologies for measuring social value. Social Value Portal claims that it provides the most reliable approach to reporting and recording social value as it is underpinned by current data and has been created by social scientists. The TOM system is designed to be tailored to particular needs, locations, sectors and KPIs, and is aligned with the UK Government's Social Value Model. The creators of the TOM system claim that social value can be more robustly calculated with their system, which allows organisations to understand their performance and progress against their aims, objectives and targets. This can aid decision-making and drive up efficiency. The TOM system is derived from the social value themes of Community, Work, Economy and Planet. It was created to avoid double-counting in reporting the same target twice. It can also assist organisations to undertake commitments to social value which are configured and relevant to their businesses. Furthermore, the system will only consider those elements which offer additionality and are beyond the legal and contractual requirements of business as usual for organisations.

7.7 Summary

Social value measurement attempts to comprehend and record the relative importance that individuals, groups or organisations place on changes relating to well-being that they experience. Measuring social value is important as businesses can then have a good estimation of the impacts that are brought about through their interventions and decision-making. It should involve key individuals who may be stakeholders from different areas of the business. These individuals will have different reasons for wanting to quantify social value through measurement. Notwithstanding this premise, all stakeholders will want assurances that their money is being used productively and effectively, achieving value for money and generating beneficial and tangible real impacts.

Social value is concerned with those outcomes which potentially change people's lives and improve their well-being, rather than just outputs linked to initiatives which may have no longer-lasting effects on individuals or communities. For this reason, social value assessment revolves around the measurement of the wider outcomes that are directly attributable to the actions of an organisation. These outcomes are sometimes referred to as 'soft outcomes' owing to the challenges faced with quantifying and measuring them.

Measuring social value can bring many benefits and allows organisations to justify spending and their investments into a particular area by looking at the impact this has had. In this way, it gives them the ability to defend their roles and decision-making in providing the services they have undertaken. There are many different ways and approaches to measure social value, including cost-benefit analysis, social accounting techniques, social return on investment and basic efficiency resource analysis.

Cost-benefit analysis is financially based and involves looking at the expected projected benefits against the costs associated with their implementation. Conversely, the social accounting techniques involve generating both quantitative (numeric) and qualitative (descriptive) data relating to the organisation's aims and objectives. The social return on investment method places a financial value on the wider impact that organisations make on social value, using what is referred to as a 'multiplier effect'. This method of transporting wider benefits into financial outcomes proved particularly beneficial in justifying and obtaining funding from both public and private sectors. The basic efficiency resource analysis is intended to generate a concise framework for evaluating activities, campaigns or programmes which are complicated and challenging. It measures the impact compared to the resources required to generate the impact. Furthermore, it looks at how efficiencies can be generated to help organisations improve their ongoing performance levels.

There are alternative ways to measure social value, and the 'TOM System' is one that encapsulates Themes, Outcomes and Measures into its methodologies. It is designed to be tailored to particular needs, locations, sectors and KPIs, and is aligned with the UK Government's Social Value Model.

Reference

Saxon, R. (2005). *Be Valuable: A Guide to Creating Value in the Built Environment*. London: Constructing Excellence.

8

Communicating Value

8.1 Introduction

In an increasingly competitive and value-driven marketplace, the ability to communicate value is just as critical as creating it. Whether bidding for a public sector project or marketing organisational achievements, it is often the organisation that articulates its value most effectively, not necessarily the one that creates the most impact, that is the most successful. This chapter explores why, how and to whom value should be communicated, and what happens when it isn't communicated effectively. Clear, honest and targeted value communication ensures all stakeholders understand the benefits delivered. This involves identifying all stakeholders through mapping and analysis, tailoring messages to their needs and selecting appropriate channels and communication outlets.

The purpose of value communication lies on a spectrum: from self-promotion aimed at winning work to altruistic knowledge sharing that uplifts industry standards. Most organisations operate somewhere in between. Authenticity plays a pivotal role. Stakeholders will only perceive communications as genuine when they align with the organisation's purpose and actions. Yet, communicating value is not without risk. Errors in measurement, poorly chosen communication platforms or mismatched communications (between stakeholder preferences and organisational delivery) can backfire. The most damaging, however, is greenwashing, misrepresenting value, particularly environmental or social, for reputational gain. This practice not only undermines trust but threatens the integrity of genuine value-driven organisations.

To be effective, value communication must be clear, simple, honest and authentic. Organisations should recognise that value isn't just in extraordinary projects, but also in everyday, beneficial practices, provided these are truthfully and thoughtfully conveyed.

This chapter offers a practical guide to ensuring value creation is not just impactful, but recognised, understood and celebrated across all stakeholder groups.

8.2 The Importance of Communicating Value

Once any type of value is created, at any stage of work, by any stakeholder, it is of paramount importance that such value is then communicated effectively. For example, if two organisations are both competing for the same public sector project, it is often the organisation that best articulates and communicates its value that is successful in the procurement. Often, even if the unsuccessful organisation is the one that actually created the most value. This is not for any untoward or unethical reasons (we'd like to think), but simply because the procuring organisation (and any advising consultants) can only mark what is in front of them. If an organisation does create impactful value but cannot communicate such value and impact effectively, how is the client to know and understand this? Therefore, the importance placed on value communication should arguably be equal to the importance placed on value creation itself. Often, however, this is not the case. This is unfortunate for both any organisation that creates impactful value but is unable to effectively articulate it (as they will fail to win work) and also unfortunate for any would-be value recipient, as the value that would be created for their benefit (if the organisation is successful at procurement) is now not created.

8.3 Value Communication Methods (Stakeholder Identification)

There is a wealth of methods that can be utilised to communicate value. Studies have shown that forms of communication of value-driven behaviour can range from annual formal summaries linked with financial reporting or ad hoc informal messages (Thorne et al., 2014). The selection of the most appropriate method actually largely depends upon the targeted stakeholder group. It is therefore imperative that a process of stakeholder mapping and analysis occurs prior to any planned communications. Through the identification of all stakeholders, each can be matched with the most appropriate value communication as well as the most appropriate value communication method. It is of little benefit to target one stakeholder group via a communication method they do not use or understand. Figure 8.1 illustrates a potential stakeholder map for a construction main contractor.

It is always imperative to conduct stakeholder mapping and analysis prior to communications occurring, so that the communications can be tailored to meet the needs of each stakeholder group as identified in the analysis. Table 8.1 shows the potential needs of the identified main contractor stakeholders.

Once the stakeholders have been identified and their needs revealed, appropriate communication methods and messages can be planned. For example, for the stakeholder group of 'previous clients', their need to 'being reassured of previous procurement decisions and existing projects' can be achieved by the contractor through articles published in the trade press. Such articles highlighting current best practice in a good level of detail will allow previous clients to feel satisfied in their prior procurement decisions, happier with any buildings they now use that were completed by the contractor, and more likely to consider using the contractor in future and recommending them to other potential clients. Similarly, the stakeholder group of 'members of the public' and their need for 'general positive sentiment' can be achieved through smaller, more superficial statements and summaries, with links back to more robust articles available on the organisation's website to provide further

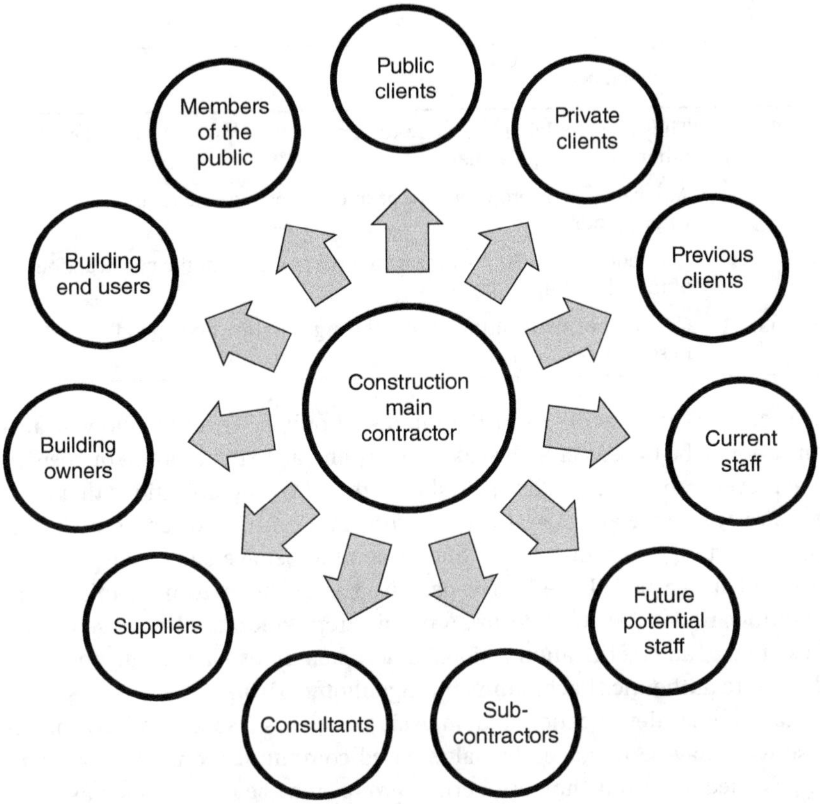

Figure 8.1 Construction main contractor stakeholder mapping.

Table 8.1 Main contractor stakeholders' potential need.

Stakeholder	Potential need
Public clients	Confidence that future procurement decisions that value can be delivered.
Private clients	Reassurance of a value-driven contractor to avoid future embarrassment and project failure.
Previous clients	Being reassured of previous procurement decisions and existing projects.
Current staff	That they work for a value-driven organisation that looks after their employees and wider society.
Future potential staff	They will work for an organisation that delivers value to society.
Sub-contractors	Higher confidence in the operations and impacts of the main contractor they will be working for.
Consultants	Trust that the main contractor will deliver the value required and expected.

(Continued)

Table 8.1 (Continued)

Stakeholder	Potential need
Suppliers	They are supplying to a value-driven organisation that will not exploit others in the supply chain.
Building owners	Assurance their projects have been completed to the required value standard.
Building end users	Awareness of the high-value standards achieved in the construction of their building.
Members of the public	General positive sentiment that the organisation does 'good' in society.

reading and resources for those members of the interested public. This will allow awareness of the organisation to be raised across broader society and a greater amount of positive sentiment to be achieved generally. Studies have shown that those organisations that have a greater general positive image are more likely to withstand future indiscretions (Saeidi et al., 2015). Whilst most organisations do not often intend to behave badly (purposefully or accidentally), it will always be of benefit to be perceived as a value-driven organisational in case anything untoward does occur in future. A positive reputation can be utilised to fall back on, and it also makes any rectification actions undertaken in response to any indiscretions be viewed as more authentic (Bachmann and Ingenhoff, 2016).

Conducting a stakeholder identification and analysis exercise is also key to fully understanding all those who may be impacted by value-based communications. Without consideration being provided to all stakeholders, certain groups may be missed, such as those indirectly impacted by organisational decisions. Missed stakeholders may lead to missed communication opportunities, which can then lead to missed benefits being realised in a timely and effective manner.

8.4 Why Value Is Communicated

Cynically, it could be claimed that communicating value is a self-indulgent process purely aimed at raising organisational awareness amongst clients with the sole aim of winning future work. Altruistically, the communication of value could be classed as a means of promoting best practice to share value creation lessons with fellow organisations and wider society. The truth is that valuable communication is often somewhere in between these extremes. Along a continuum that fluctuates from work-winning image-driven self-marketing to a true collaborative and sharing of lessons learnt, driven by altruistic motives. Whilst the preservation of commercial advantage cannot ever be underestimated, the prevailing discourse is now one of creating value and sharing how it is created for a wealth of diverse benefits across multiple stakeholder groups. The why of value communication, therefore, falls along the continuum illustrated in Figure 8.2.

Organisations should communicate all value they create (even if there is a delay between the value creation and the value communication to maximise any commercial advantage). Often, an organisation may not think about where on the continuum

Figure 8.2 Value communication continuum.

their value-based communications occur but having a greater awareness of the purpose of each communication, the communications can be better positioned to serve their intended purpose. This knowledge, combined with stakeholder identification and analysis, will serve to further the effectiveness of all value-based communications. However, often an organisation may intend for its initiatives and communications to be altruistic, but if they are not perceived as such, and perceived as inauthentic (Section 8.7), then stakeholders may view any communications as purely self-serving and not truly reflective of an organisation's intentions. This could then lead to the value an organisation is creating, or intends to create, being undermined, and accusations of greenwashing (Section 8.10) may be applied.

8.5 How Value Is Communicated (the Methods)

Once the stakeholders are identified and the purpose of the value-based communication is understood, the methods of communication can be ascertained. This was touched upon previously but is developed further below.

There is a plethora of communication methods available, and a contemporary organisation should look to maximise all available outlets as part of its ongoing marketing and communications strategy. Not least because, as societal views evolve and geopolitical changes occur, some communications platforms may fall out of favour and association with them may be viewed either positively or negatively. Therefore, to be active across several platforms will allow an organisation to both reach as wide a variety of stakeholders as possible and be able to quickly pivot their communication strategies from one platform to the next, being more resilient if they decide to stop utilising one platform in response to societal perception changes.

The points below are examples of the high-level communication choices and platforms that may be utilised by an organisation in communicating value.

- Blogs on the organisation's own website
- Press releases on the organisation's own website
- Social media posts
- Professional networking posts
- Speeches at events and conferences
- Articles in trade press
- Newspaper and magazine advertised or sponsored posts
- Green papers/White papers
- Tender submissions
- Annual reports
- Conference/Journal articles
- Textbooks and non-fiction book releases

8.6 What Is (and Is Not) Value Communication

It is important for organisations to understand, however, exactly what is and is not, value communication. Having many different value disciplines available results in numerous topics, concepts and general ideas being discussed. Add to this the numerous definitions and different understandings and interpretations of each sort of value, and the opportunity for misunderstood and incorrectly received communications is high. Therefore, it is of paramount importance that the individuals and organisations communicating value are clear on exactly what value is.

Contrary to some beliefs, value can be the benefits that arise from everyday business-as-usual practices. Indeed, many organisations that are effective at value communication and not the ones that create the most impactful value. They are often the ones who are just better at capturing the value they create, and better at articulating and communicating that value to the intended stakeholder groups. Therefore, in simplistic terms, value can be (and is) anything and everything that is of benefit to the natural, human and built environment that derives from an organisation's actions. This can be direct and indirect; it can be immediate and long-term and it can even be intended and unintended.

Value is not strategies, actions and impacts that are created in isolation from an organisation's actions. Such things may result in value, but in such instances, it would not be value attributed to organisations not involved in their creation. Value is also not what would happen anyway. This one is trickier to understand and accurately measure, but if an action would always have occurred with or without an organisation's involvement, that organisation should not claim the value creation as their own.

8.7 Authentic Value Communications

Authentic value communication sits hand in hand with authentic value creation. Being authentic can be described as staying true to one's own principles and undertaking initiatives that align with an organisation's actions, impacts and general purpose. It is about undertaking initiatives that are both positively impactful for society and address concerns that are important and linked to the operations of an organisation. Authenticity is perceived by an organisation's different stakeholder groups and so is not applied or awarded by the organisation itself. An organisation cannot simply say it is authentic; such authenticity is either awarded to it or not, depending upon how such actions are perceived by stakeholders.

However, an organisation can increase the chances of its actions, strategies and communications being viewed as authentic by ensuring all value behaviour aligns with the organisation's purpose and existing markets and operations. For example, authentic value could be where a developer is planning to build over land that houses existing trees, which will be removed to allow the development to occur. Environmental value could then be created by the developer planting replacement trees in the same area at a ratio of 3:1. Therefore, for every tree that is removed as part of the development, three trees will be planted in its place. This ratio is not fixed and could be amended upwards or downwards depending upon the maturity of the trees removed and the types of trees planted in their place.

One issue that could arise (theoretically at least) is the awarding of value by one stakeholder group but not by other stakeholder groups. In such an instance, the organisation would have to prioritise the different stakeholder groups and ensure their focus and efforts were dedicated to achieving legitimacy from the group(s) they rank as the most important. However, in practice, doing so (focusing on one stakeholder group at the expense of others) would likely lead to a loss of perceived authenticity from all stakeholder groups. The marginalisation of one stakeholder group would inevitably lead to discontent spreading amongst all stakeholder groups. For example, if current and future clients found out that a main contractor showed no awareness of the value concerns of previous clients, it would undermine any value-based communications the main contractor was aiming at current and future clients. Equally, if a main contractor showed no concern for the value expectations of building end users, any other value-based initiatives and communications would not be viewed as authentic and would be perhaps considered more self-serving than altruistic on the value communication continuum. It is therefore imperative that organisations understand the needs of all stakeholders. Whilst value-based communications can be aimed at certain groups and with certain outcomes in mind, no stakeholder group should be ignored with regard to value-based communications and initiatives.

8.8 Ambiguous and Precise Communications

A central tension in values-based communications revolves around whether organisations engage with value behaviour substantively, through genuine, meaningful action or symbolically (through greenwashing). The latter refers to organisations presenting an image of responsibility that does not reflect their actual practices. Research suggests that symbolic engagement is widespread, making it increasingly difficult for stakeholders to determine whether value-based communications reflect authentic efforts or mere rhetoric. This ambiguity leaves room for diverse interpretations and creates a potential disconnect between organisational intentions and stakeholder perceptions. Crucially, the role of ambiguity in values-based communication is not merely a flaw or limitation; it can also be a strategic choice. Strategic Ambiguity, as conceptualised by Eisenberg (1984), provides a useful lens for understanding how ambiguous language can serve organisational purposes. Strategic Ambiguity enables organisations to present information in ways that are deliberately open to interpretation, thus allowing them to appeal to multiple, often conflicting stakeholder groups simultaneously. This is particularly relevant in value-based communications, where organisations must address a wide array of audiences, including investors, consumers, regulators and advocacy groups, each with different expectations and understandings of what constitutes socially responsible behaviour. Ambiguous language can help organisations manage these divergent interests. It allows them to: (i) maintain flexibility and adaptability, (ii) promote multiple interpretations without committing to specific details, (iii) include broad, value-laden terms that resonate with diverse audiences and (iv) revise narratives over time without appearing inconsistent (Fernando and Sim, 2011). Wexler (2009) found that such ambiguity in CSR reporting helped to foster shared understanding and alignment, even where consensus was lacking.

However, relying on ambiguity comes with risks. Overuse can erode trust, especially if stakeholders begin to suspect that vague language masks a lack of genuine value, behaviours and performance. While ambiguous value-based communication may be effective in managing stakeholder relationships, it can also obscure poor practices and limit transparency. Therefore, understanding the interplay between ambiguous and precise language is crucial for evaluating values-based communication. Many value communications, whilst useful, must be critically analysed alongside other organisational practices and stakeholder experiences to accurately assess the authenticity and impact of all value initiatives. Ultimately, value-based communication is not just about what is communicated, but how it is both communicated and how it is understood.

8.9 The Advantages and Disadvantages Associated with Communicating Value

Whilst the advantages of communicating value have been outlined throughout this chapter, the following points recap and expand on the benefits of value communication:

- Encourages others to create value
- Increases the amount of value created across society
- Reassures previous clients/existing projects that they made the correct choice with regard to contractor selection and
- Shares best practices with others
- Helps win future work/have information readily available to include in future bids pertaining to value
- Makes a positive difference to the value recipients
- Helps reputational enhancement, which then helps protect against any future indiscretions by minimising negative sentiment
- Internal pride amongst staff can also increase at working for an organisation that creates impactful positive societal value.

The following points recap and elaborate on the disadvantages that may be associated with value communication:

- It can be easy to make mistakes when collecting data and calculating values that may lead to incorrect communications.
- Any mistakes made in one area may then undermine correct claims made in another area.
- It could lead to reduced competitive advantage if competitor organisations then use the same initiatives.
- Any embellishments made may then be subject to claims of greenwashing.
- All backup data needs to be readily available to withstand any scrutiny.
- It can be quite resource-intensive to manage all communication channels effectively on an ongoing basis.
- Value-based messages communicated to incorrectly identified stakeholder groups or via incorrect methods may not resonate as effectively as intended.

8.10 How to Communicate Value Effectively

The following points each identify and address a different theme that should be followed for all value-based communications:

- Clarity – All communications need to be clear. Having too many messages or too much information contained can distort the intended message and obfuscate any benefits described.
- Simplicity – All communications need to state the intended message in as few words as possible. A simple message is one that is remembered and often one that is received as intended.
- Honesty – It goes without saying that only honest messages and statements should be communicated. Any claims of greenwashing can cause negative reputational issues.
- Authentically – All communications should be authentic. Inauthentic value-based communications can raise questions amongst stakeholders as to the motives of the organisation, even if the motives are altruistic.

8.11 'Greenwashing' in Social Value Communication

Many organisations communicate their 'green' behaviours. These are those behaviours that substantively engage with environmental and broader sustainable items. Organisations that do this often (and rightly so) receive plaudits from stakeholders for such communications with research showing clear links to enhanced work winning and stakeholder satisfaction (Watts et al., 2019). However, for the longest time, questions have been asked as to whether organisations purposefully greenwash or, where occurrences are revealed, are they just from organisations not substantially engaging with value-based behaviour, yet communicating it as if they are through having a confused understanding (Michelon et al., 2015). Such debates are still ongoing and show no signs of being resolved. Ultimately, organisations will always communicate the value they believe they are creating. If commercial measurement tools are happy to be exploited to support organisations in their work-winning attempts, value behaviour will also end up being 'stretched' in how it is quantified. Previous studies have illustrated this by finding that an organisation's reported values behaviour may not in fact match its actual behaviour undertaken (Font et al., 2012).

Greenwashing itself is a term that describes where such formal communications are used by an organisation to falsely portray itself as one that substantively engages with green issues, but in fact, such communications are incorrectly used to present a false narrative and image of the organisation. It is where an organisation seeks to gain the advantages a positive green image will yield but without putting in any of the hard work required, and without actually making positive environmental or sustainable impacts. As stakeholder and wider societal expectations have evolved over time from a focus on environmental value to one of sustainable value, before focusing on social value, and now evolving to a more total value focus, the need to communicate any practices undertaken is growing more important. The benefits associated with communication of such practices are also increasing (such as the difference between successful and unsuccessful tender returns). As organisational costs rise and competition increases, all businesses are looking at how to win more work, delivered more efficiently, for as small a resource investment as possible. Therefore, the risks of some organisations adopting 'greenwashing' practices across all total value elements are increasing.

For those organisations that do greenwash their value and claim to deliver more impact and value than they actually do, this serves to undermine the whole concept of value creation. It will directly lead to fewer benefits being achieved for society and fewer positive recipients of value creation. It will also indirectly lead to all organisations delivering less value to society and to those in need. If an organisation greenwashes their practices and then is successful in winning work, without a client who rigorously checks all value created and holds the organisation to account, the organisation (and others like them) could win more and more work. Therefore, the organisations that do create value and ensure it is impactful but are losing out to others who adopt greenwashing practices, will inevitably not be able to continue operating. This will, therefore, lower both the current and future amount of value created as a whole across sectors and industries.

One study found that greenwashing, through the use of strategic ambiguity (the process of purposefully using ambiguous language to obfuscate your true behaviours and intentions) allowed contractors to circumvent the need to attend to value behaviours authentically, and could instead hide behind ambiguous word use and rhetorical statements and still receive the same benefits as organisations that authentically embrace value driven behaviour yet are restricted to communicate such behaviours in the same way (Watts et al., 2019). However, the study does show that such strategically ambiguous 'greenwashing' communications can be identified by clients through the requesting of specific information and robust reviewing of any value-based communications.

8.12 Conclusion

Communicating value is as essential as creating it, especially in today's competitive, value-driven environment. Success often hinges not on the scale of impact alone, but on an organisation's ability to clearly and authentically express the benefits it delivers. Effective value communication ensures stakeholders (both internal and external) recognise and appreciate these contributions. This requires careful stakeholder mapping, message tailoring, and selection of appropriate communication platforms. Organisations must strike a balance between promoting their achievements and contributing to broader industry knowledge. Regardless of where they sit on this spectrum, authenticity remains the cornerstone of credible value communication. When messages align with an organisation's actions and purpose, they resonate. However, missteps, such as unclear messaging, poor platform choices or inaccurate measurements, can undermine efforts. The most damaging of these is greenwashing, which not only erodes trust but risks discrediting organisations genuinely committed to delivering environmental or social value. Ultimately, this chapter reinforces that meaningful value exists not only in high-profile projects but also in consistent, everyday practices, so long as they are communicated truthfully. By applying the principles and strategies outlined here, organisations can ensure that their value is not only created but also recognised, understood and celebrated across all relevant stakeholder groups.

References

Bachmann, P. and Ingenhoff, D. (2016). Legitimacy through CSR disclosures? The advantage outweighs the disadvantages. *Public Relations Review*, 42: 386–394.

Eisenberg, E. (1984). Ambiguity as strategy in organisational communications. *Communication Monographs*, 51: 227–242.

Fernando, M. and Sim, A. (2011). Strategic ambiguity and leaders' responsibility beyond maximising profits. *European Management Journal*, 29: 504–513.

Font, X., Walmsley, A., Cogotti, S., McCombes, L., and Hauser, N. (2012). Corporate social responsibility: the disclosure-performance gap. *Tourism Management*, 33: 1544–1553.

Michelon, G., Pilonato, S., and Ricceri, F. (2015). CSR reporting practices and the quality of disclosure: an empirical analysis. *Critical Perspectives on Accounting*, 33: 59–78.

Saeidi, S., Sofian, S., Saeidi, P., Saeidi, S., and Saaeidi, S. (2015). How does CSR contribute to firm financial performance? The mediating role of competitive advantage, reputation and customer satisfaction. *Journal of Business Research*, 68: 341–350.

Thorne, L., Mahoney, L., and Manetti, G. (2014). Motivations for issuing standalone CSR reports: a survey of Canadian firms. *Accounting, Auditing & Accountability Journal*, 27(4): 686–714.

Watts, G., Fernie, S., and Dainty, A. (2019). Paradox and legitimacy in construction: how CSR reports restrict CSR practice. *International Journal of Building Pathology and Adaptation*, 37(2): 231–246.

Wexler, M. (2009). Strategic ambiguity in emergent coalitions: the triple bottom line. *Corporate Communications: An International Journal*, 15(1): 62–77.

Part III

Value Reflection

9

The Concept of Project Success

9.1 Introduction

This chapter will start by outlining the challenges for project success in the global construction industry, including the fragmented and temporary nature of the sector, and different government regulations, cultures and behaviours across the world. To respond to these challenges, will explain why many attempts have been made to improve the current situation and provide a more robust and stable environment for the industry. The chapter will then attempt to outline what constitutes project success with reference to the traditional metrics of cost, time and quality. It will then introduce the notion of performance and the different definitions and meanings around this. Thereafter, the chapter will discuss the challenges faced by the construction industry in achieving and measuring project performance and success, especially given the alternative perspectives of different project stakeholders.

The alternative approaches and models to measure project success will then be discussed from the perspectives of project efficiency, impact on customer, business success and preparation for the future. Key performance indicators (KPIs) will then be introduced and their role in assessing project performance around success. The chapter will then discuss the concept of critical success factors (CSFs), particularly around the four categories of comfort, competence, commitment and communication. The utilisation of such factors will then be examined.

The chapter will discuss the issues, dilemmas and challenges associated with measuring performance for construction sites. These factors will be reviewed from different angles and perspectives around quality assurance, site supervision and safety.

9.2 Challenges for Project Success in the Global Construction Industry

As previously referred to in the book, there are many different challenges facing the construction industry. These include the bespoke nature of the industry, where teams come together at relatively short notice, undertake a given project and then disband. Learning from one project to the next is very limited and hampered by an insufficient degree of information as to what went wrong on projects and how to correct and address these in the future. In addition, different governmental regulations, cultures and behaviours across the world can dictate how projects are procured. Whilst there has been a concerted move away from accepting tenders solely at the lowest price in modern works, some countries still condone and regulate these practices.

These characteristics can adversely affect the success of projects, especially those teams assigned to project management, where there is sometimes little scope for inter-organisational relationship development and partnering philosophies. For this reason, there have been many attempts to improve the current situation and provide a more robust and stable environment for industry. This is designed to create a more collaborative, longer-term basis for construction project management to operate within, based on the partnering principle and harbouring trusting relationships (Challender et al., 2019). Notwithstanding this premise, some project managers still adopt traditional practices and behaviours which, at times, could be described as adversarial and predicated on win-lose scenarios. Furthermore, in some cases, projects may be managed in an ad hoc and intuitive manner, with resources being allocated to them across many different areas and stages. Clearly, this could introduce unnecessary and avoidable risks in not having a more systematic and controlled basis in place, with established procedures, governance and strategies deployed. To address these inadequacies, the notion of project success needs to be more carefully considered from the outset by all stakeholders, including project managers. However, what constitutes 'success' is not always a constant and is open to debate and interpretation amongst construction professionals and other project stakeholders, including clients. For this reason, Chan (2001) developed a framework for the measurement of success on projects in the construction industry. In this pursuit, a range of KPIs was created as a benchmark for the measurement of construction projects from the perspective of performance. These factors covered a combination of planned and unplanned interactions and events which affect projects at all stages of their procurement and development over the project life cycle.

9.3 What Constitutes Project Success?

There are numerous definitions of success. According to Arslan and Kivrak (2008), success is the extent to which expectations and goals are achieved. Conversely, they asserted that the definition of failure is the inability of a business to carry out its agreed obligations and commitments.

Notwithstanding the above definition, what constitutes success on construction projects is not always universally the same across all viewpoints. This is partly attributable to the complex and dynamic nature of the construction industry. The industry has created an environment which is dynamic and ever-changing, partly due to ever-growing uncertainties in areas such as budgets, technologies, procurement and design and development processes. In addition, construction projects are delivered within a labyrinth of different environments and contexts, and there is a vast array of stakeholders who have varying opinions and beliefs as to what success means. Accordingly, individuals, groups and organisations will have different ways to judge and measure success, and for this reason, successful outcomes for projects mean different things to different people. The 'iron triangle' of cost, time and quality, as advocated by Atkinson (1999), is commonly referred to as the three success factors for projects but some would regard this as too simplistic in what is described as a complex phenomenon. Accordingly, there are many other determining factors of what constitutes success. Furthermore, stakeholders will usually change throughout the course of every project, as projects transgress through the various design and construction stages. In addition, the environment within which they are procured will normally be constantly changing (Sanvido et al., 1992).

9.4 The Meaning of Performance

There are many different definitions and meanings around what constitutes performance. Early definitions and descriptions focused on the concepts and theories linked to behaviour and motivational practice. Vroom (1964), for instance, advocated that both ability and motivation are products of performance. Conversely, other perspectives have emerged in the past on what constitutes performance. Other examples have created theories which have associated performance with efficiency, satisfaction and effectiveness. Notwithstanding the above different views, there cannot be one universal understanding of what constitutes performance. This assertion is predicated on the premise that performance means different things to different organisations and stakeholders. According to Salminen (2005), the measurement of performance can be through 'soft' metrics, such as employee attitudes or 'hard' metrics, which include financial and non-financial metrics. In the latter case, this could encapsulate sales, revenue, profit levels, as well as customer satisfaction and repeat business data. Salminen (2005) also opined that performance covers both results and processes and can encapsulate most things which underpin or describe the success of an organisation. Performance, however, should be differentiated by other factors such as decision-making and business strategies, which can generate success themselves. To understand this, organisations could be reaching high levels of performance but still be unsuccessful owing to poor business strategies being implemented. This clearly presents another complication and dilemma on what performance means in practice.

9.5 What Are the Challenges Faced in Achieving Project Success and Performance?

It is essential for companies to remain successful to maintain their survival within competitive business environments. The construction industry operates within a highly dynamic environment characterised by changes brought about by new technologies and business methods. This can present a dilemma for those construction-related companies, large and small, in being able to adapt to these ever-changing challenges and develop appropriate strategies to become more competitive (Arslan and Kivrak, 2008).

9.6 Measuring Project Success

There are many different perspectives and methodologies for measuring project success. This section of the book will give an overview of the various schools of thought around this specific subject area to help readers identify those criteria and elements which are more important than others.

9.6.1 The Concept of Assessing Project Success

The concept of 'project success' can be construed as abstract, with many complex and conflicting factors and determinants which could determine whether projects are successful or have resulted in failure. Consequently, Chan (2001) developed their framework for evaluating project success through a comprehensive review of previous research and literature from numerous sources. From the UK, these included:

- Engineering, Construction and Architectural Management
- Journal of Construction Procurement
- International Journal of Project Management
- Construction Management and Economics

The research undertaken by Chan (2001) identified many different definitions, criteria concepts of what constitutes project success. For example, the achievement of defined objectives or specific favourable outcomes was commonly regarded as targets for success, but others related to standards of principles and judgements measured for value. Notwithstanding these assertions, it is generally accepted that there is no universal acceptance of what success means, as this is open to different interpretations from different groups of people and therein unequivocal and subjective. For example, architects may regard design and aesthetics as the main factors for success, whereas clients and end users may regard operational functionality and delivery of projects on time and budget as more important. In addition, views on success can be affected by the varying nature of projects, dependent on their size, value, complexity and use. These different factors may become the criteria that are used to decide if they go on to win awards and thus 'differentiate them from the pack' as exceptional projects. Conversely, they may be simply regarded as 'run of the mill' building projects and consequently denigrated to this lower tier of status and significance. Perhaps, therefore, it is not surprising that architects are constantly seeking those projects which will reach higher accolades, building reputational benefits for their practices and assisting them in winning future contracts.

Measures for project success, according to Pinto and Pinto (1991), should not be restricted to completed buildings but should also include outcomes based on psychosocial factors. These include the relationships developed within project teams and the overall level of satisfaction and contentment of the participating team members. These relationships and the satisfaction that comes with them could be characterised as 'soft' measures. This presents a somewhat conflicting perspective in that projects could be delivered late, over budget and to a quality standard below expectation, but could be deemed successful if teams have maintained good working relationships over time and while completing projects.

Another determining factor for success could be related to maintaining health and safety throughout projects. It is not unreasonable to deem those projects where there have been accidents, injuries or even fatalities as unsuccessful. This is especially the case if there are circumstances which have been found to have breached regulations or established safety standards. In the UK, the Health and Safety Executive (HSE) will investigate incidents where reportable incidents and accidents have occurred on construction sites. In some cases, where breaches have been identified, this can lead to prosecution of those organisations or individuals involved. In addition, breach of health and safety can lead to other project complications, including delays as work may be suspended, and claims for financial

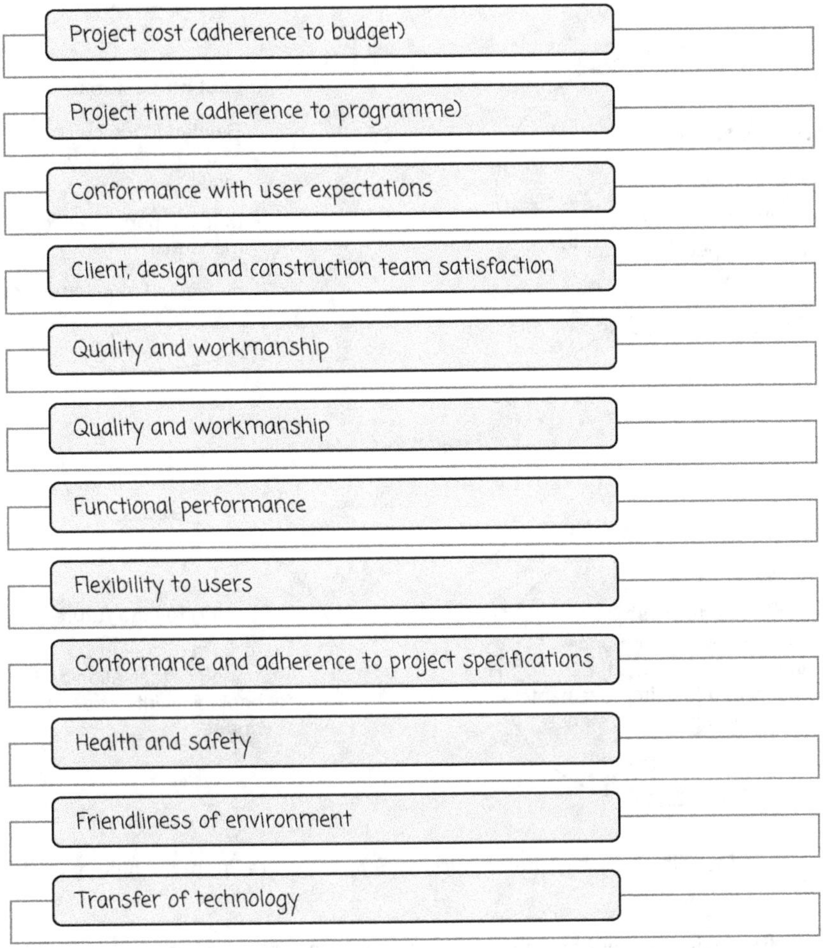

Figure 9.1 Various success factors. *Source*: Adapted from Chan et al. (2001).

losses and litigation proceedings follow. From the research undertaken by Chan (2001), the various success factors could include those depicted in Figure 9.1.

Perhaps what is missing from the above list, which has become a more important focus since the research of Chan (2001) took place, is the growing agenda for achieving environmental sustainability. This is especially important given climate change and the global focus on reducing carbon emissions. In the UK, there is a growing impetus on delivering projects which achieve levels of environmental sustainability and attain accreditation from schemes such as BREEAM and Passivhaus (Challender and Challender, 2024).

9.6.2 Alternative Approaches and Models Assigned to Measuring Project Success

Shenhar et al. (1997, as cited in Chan, 2001) took a different approach and advocated that there are four aspects or dimensions to project success. These are shown in Figure 9.2 and related to project efficiency, impact on customer, business success and preparation for the future.

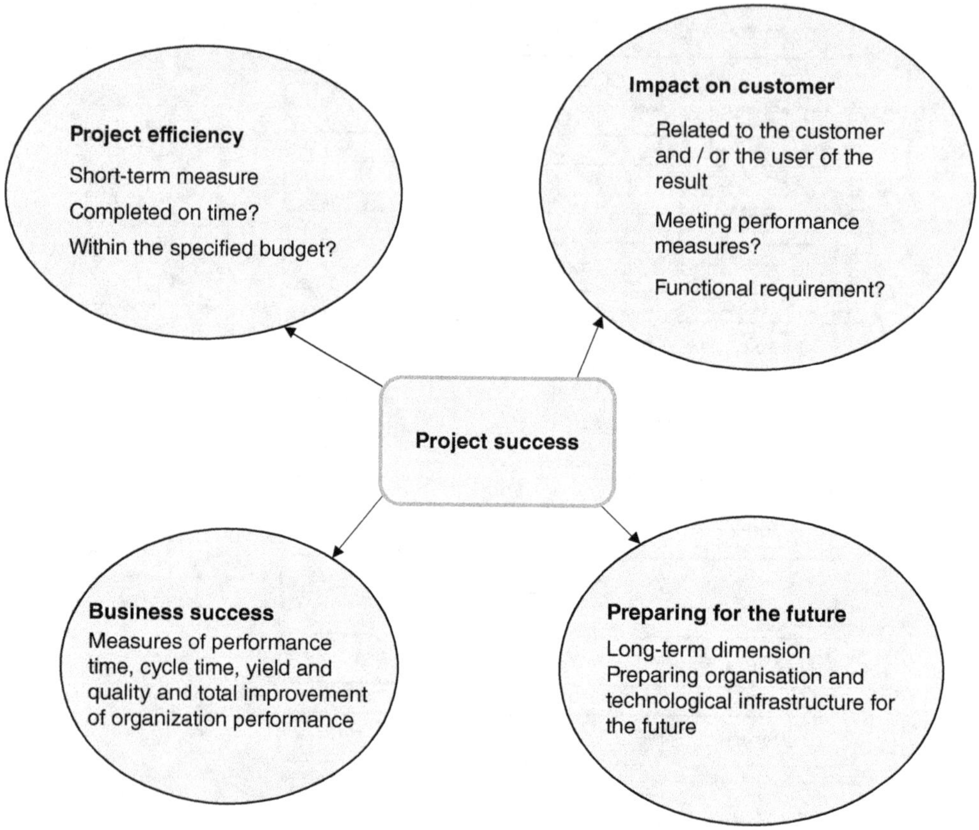

Figure 9.2 The four dimensions of project success. *Source*: Adapted from Shenhar et al. (1997).

The first dimension, namely project efficiency, is related to the contract period timescale during the execution stage of projects, whereas the second dimension, impact on customer, is within the context of a short time post completion, after buildings are delivered to their clients/owners. The third dimension, business success, is related to a longer period after completion, possibly one to two years, when the level of business success attributable to the project, including possible sales and profits, can be established. Finally, the fourth dimension, namely, preparing for the future, deals with a period possibly three to five years post completion, when longer-term factors, including preparing for further changes, can be assessed.

In a similar way to Shenhar et al. (1997), Atkinson (1999) developed a three-stage model for articulating project success. This was predicted in a delivery stage where evaluation of time, cost, efficiency and quality could be determined. In addition, within a post-delivery stage, there were two stages which represented both the system and the benefits of getting projects right. The former related to benefits to project stakeholders and the procurement system to deliver the projects, whereas the latter was concerned with the impact on customer and business success.

Sadeh et al. (2000) also devised a four-stage dimension model for project success. The first dimension related to achieving design goals and measured by evaluating time, cost,

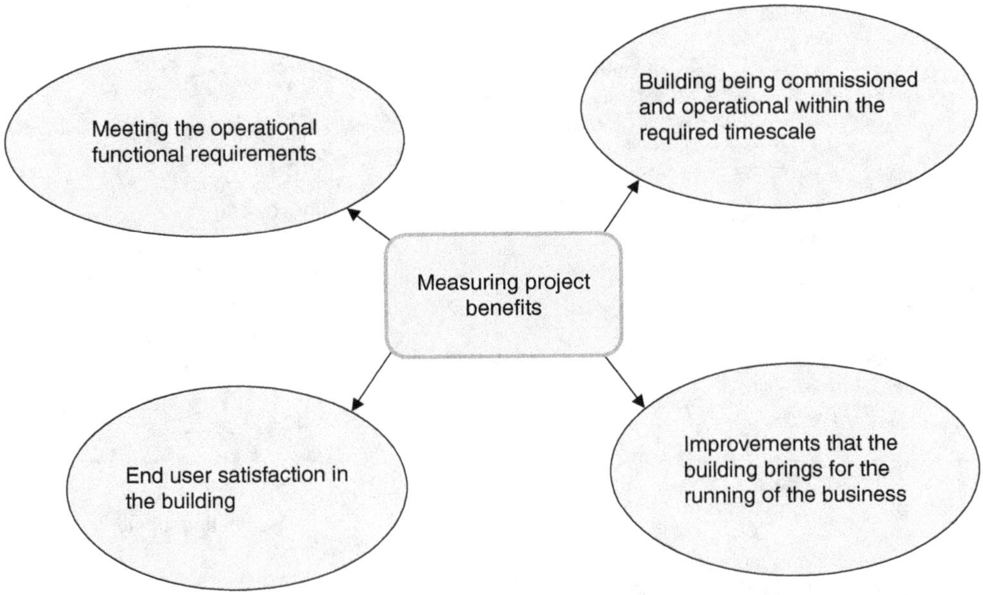

Figure 9.3 Factors for measuring project benefits. *Source*: Adapted from Sadeh et al. (2000).

functional and technical performance. Conversely, the second dimension relating to attaining benefits for end users could be measured from the following factors in Figure 9.3.

The third dimension relates to direct benefits brought to the organisation, including increasing profits, introducing new markets and products, growing technological capabilities and gaining an improved reputation in the marketplace. The final dimension related to the benefits associated with technological infrastructure brought about by projects. This included decreased dependencies on external sources, contribution to other projects and a flow of information and contribution to other critical subjects and generations.

The different models described above present many different approaches to what constitutes success for projects. For this reason, Figure 9.4 has been devised to summarise the various findings on what could be defined as CSFs.

Put these into a spider diagram.

9.6.3 Key Performance Indicators

According to Chan (2001), in the context of the construction industry, KPIs are designed to measure the performance of projects and those organisations undertaking projects. In this regard, they are regarded as general indicators of performance with a focus on outputs and outcomes, which are deemed important and critical. For ease of use, the number of KPIs should be reduced to a manageable number and data collection methodologies should be as simple as possible to avoid complex evaluation. In addition, they should be designed with a generic nature that can be adaptable and utilised on many different projects. For this reason, indicators and measures should be universally accepted, owned and understood by different stakeholders across organisations and therein not specifically designed around one group or business area. For future development and adaptability, KPIs should be able to be updated, changed or refined as other possible requirements emerge.

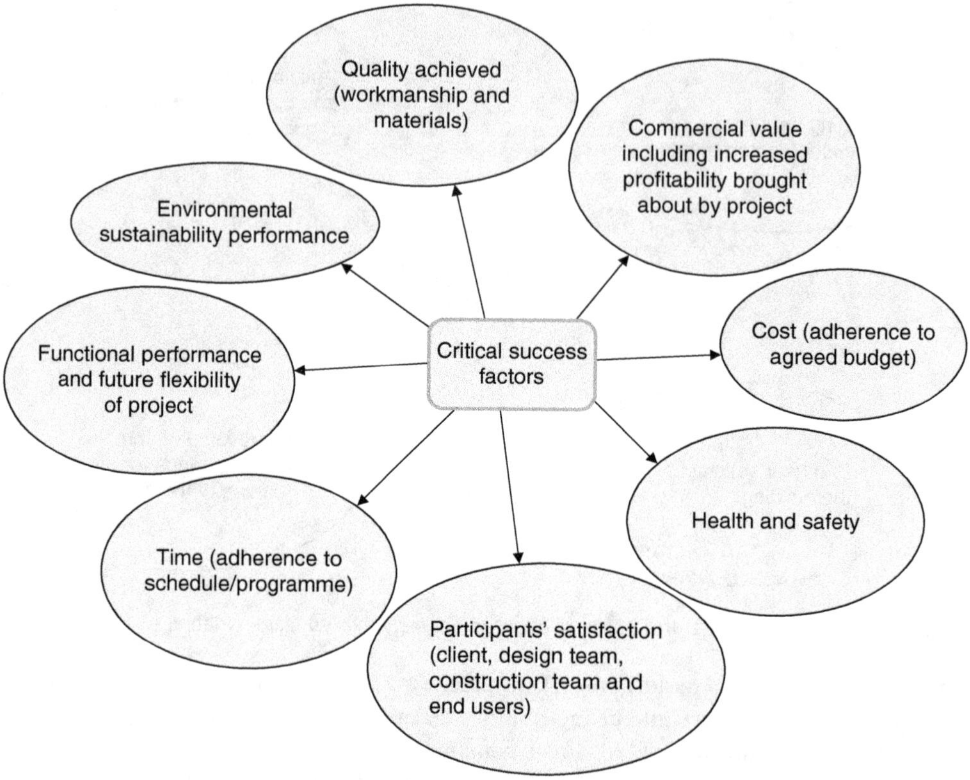

Figure 9.4 Diagram to summarise the various findings on what could be defined as critical success factors.

9.6.4 Analysis of Key Performance Indicators for Projects from Lessons Learnt Reflective Practice

Reflective practice is an important area to determine the success of projects, and this is an area that will be covered later in the book.

KPIs can be grouped into two distinct areas, around both objective and subjective measures. Objective measures can be more easily measured from tangible data, including construction costs, net present value, time variations, accident rates, environmental impact and unit costs (£/m^2). Conversely, subjective measures could include functionality, quality and satisfaction of end users, clients and project teams. Chan (2001) advocated that a seven-point scoring system could be applied against both these objective and subjective measures, ranging from very dissatisfied to very satisfied. At the end of a project, reflective practice for lessons learnt should dictate that a post-completion workshop is conducted. This should be attended by all project participants, including design and construction teams and end users and clients. It should consider some or all of the KPIs referred to above for which should be assessed by the different workshop attendees. Where applicable, key successes should be discussed and debated across the participants. In addition, the possible applications and inclusion of the attributing factors should be agreed upon for their use on future projects, to develop successive successful project outcomes. Notwithstanding this premise, to celebrate the success factors, one should also identify, interrogate and closely analyse those areas where performance could be improved.

There may have been issues, events, actions or behaviours which could have directly or indirectly led to project complications emerging. The circumstances surrounding this area should be reflected on from a positive perspective, without members of the team feeling they are being unduly criticised and where defensiveness of participants could come into play. The main aim of this reflective practice is to understand what events or outcomes did not go as well as planned and expected, rather than being simply concerned and focused on project failures. In this way, safeguards against such problems emerging on future projects could be discussed, agreed and documented. In some cases, improvement measures identified could be incorporated into future project controls, procedures and best practices to avoid repetition of similar problems arising. This notion of continual improvement through reflective practice from past experiences can assist value management, and this will be discussed more in the next chapter.

When considering the notion of success, as it is a subjective and debatable commodity, it is essential to measure actual performance against expected performance. In this way, establishing some of the KPIs previously referred to and having mechanisms in place to fairly and robustly measure the scale of success is essential for evaluation and future learning. This can then establish helpful information and create a benchmark for future development programmes and research.

9.7 Critical Success Factors Influencing Project Success in the Construction Industry

9.7.1 The Concept of Critical Success Factors

CSFs can be defined as specific action areas or elements where a team, business or department should focus on to achieve their successful implementation and therein attain strategic objectives. The successful execution and implementation of these success factors should deliver positive outcomes and generate meaningful value for an organisation.

9.7.2 Identifying and Utilising Critical Success Factors

Garbharran et al. (2012), referencing Nguyen et al. (2004), identified four categories of CSFs. These were referred to as the 'four COMs' and comprised communication, competence, comfort and commitment.

1. **Comfort**
   The comfort category emphasised that stakeholder engagement is an essential requirement for executing successful projects. Garbharran et al. (2012) explained that stakeholders in this regard could be classified as primary or secondary. Primary stakeholders are those who hold a legal connection with projects, and examples of these could be clients, design team consultants, main contractors and sub-contractors. Conversely, secondary stakeholders do not have a contractual or direct relationship with projects but are nevertheless influential decision makers. These could include local authorities, community groups and funding bodies. Garbharran et al. (2012) stressed that the requirements of all stakeholders must be carefully considered and managed effectively if projects are going to succeed. An essential prerequisite, they explained, is to

appoint a project manager who, amongst other duties, can harness the inputs from stakeholders. To ensure appropriate competency, a project manager should possess the necessary technical skills but should also have the 'soft skills' associated with transformational leadership, team management, emotional intelligence and conflict management. There is also a requirement to ensure that there are sufficient resources allocated to projects. Accordingly, resource management plans should be formulated in close consultation with all relevant stakeholders. However, this can sometimes involve phenomena and conflicts associated with competition for finite resources from different groups. One example could be the allocation of funding for the main construction contract and additional budgetary allocations for all other expenditure, including equipment, fixtures and fittings and information technology hardware. Another consideration to avoid dilemmas is that there should be adequate resource contingency planning to cater for unexpected events and developments which may be experienced throughout the various stages of project execution. Examples could include additional costs associated with unexpected ground conditions or periods of high construction inflation, as was experienced in the UK between 2021 and 2023, where costs can increase rapidly. Garbharran et al. (2012) stressed that financial plans should be developed that itemise all elemental costs of projects and incorporate them within project activity schedules. They explained that this should ensure that projects are sufficiently funded. In addition, contract documentation must be robust from the perspective that relevant stakeholders are legally accountable and enter into binding contractual agreements. Such agreements should safeguard that they deliver activities competently, against prescriptive cost, time and quality parameters to achieve the required levels of performance in carrying out their duties.

2. **Competence**

   According to Garbharran et al. (2012), competency can, alongside comfort, be regarded as being paramount to achieving successful project outcomes and can be categorised into four component aspects. The first aspect of competency relates to the utilisation of modern technology to harness its full potential is crucial in this regard; it can give a competitive advantage for construction organisations. The construction industry has developed many different technologies over recent years, and therefore, it is important to select those technologies that can generate optimal levels of utilisation. In addition to technology, the second component aspect of competency should be an emphasis on using lessons learnt and reflective practice from past performance to encourage improved working practices for projects. This aspect will be covered more in Chapter 10, Reflective Practice Around Value Achievement on Projects. Without this knowledge learnt from previous case studies and experiences, it would be very easy to make the same mistakes and poor decision-making on future projects. Accordingly, Garbharran et al. (2012) described this as 'tacit knowledge'. They advocated that such knowledge gathered from experiences should be documented so it is not lost or forgotten about during all project stages, as therein acts as a benchmark for relevant decision-making. The third component aspect of competency relates to deploying teams who are suitably skilled, experienced and have the right expertise and knowledge for the nature of the projects they have been assigned to carry out. To ensure this is the case, there should be a comprehensive analysis of skills carried out, which should identify any deficiencies or gaps which need to be filled. The fourth and final component factor around competencies is related to the selection of the most appropriate project manager and main contractor. The selection process is crucial in this regard, using the most appropriate and relevant selection criteria for tender bods to be evaluated against. These should include aspects such as specialist experience and expertise, track record, quality assurance accreditations, technical proficiencies and health and safety credentials. In addition, where possible, references or testimonials should be sought from past clients of similar projects to gain another insight into the suitability of the bidding organisation. In no circumstances should bids be solely judged and decided upon on the lowest tender price, as this is a recipe that will generate the worst possible outcome for projects (Latham, 1994; Egan, 1998, 2002; Challender, 2022, 2023).

3. **Commitment**
It is essential to commit to all projects, especially from senior management within organisations. To ensure that this commitment is not ambiguous in any way, there should be clear and comprehensive aims and objectives established for projects. In this way, this should avoid any misinterpretation of what has been agreed upon amongst all stakeholders and committed to. A common source of discontentment on any project can occur when project teams wrongly assume that additional funding can become available at any stage in the life of a project. Such requirements for additional funding may have emerged from construction complications on site during the delivery phase, and examples could include delays from inclement weather, adverse ground conditions, availability of products and materials and price inflation. Additional funding may not always be available, and once budgets have been established and approved, any further financial commitments may be resisted by client organisations. This can sometimes lead to cost-cutting or what is commonly referred to as 'value engineering'. In fact, this process of value engineering rarely generates value in practice and is normally synonymous with reducing the scope of projects by the removal of works deemed as non-essential. Notwithstanding this process of reducing cost to meet budgetary constraints, it can have the adverse effect of compromising projects by taking away those elements that could enhance the design and functional performance of buildings and facilities, and thus reducing overall quality standards.

Notwithstanding the above issues and examples, commitment to projects is not always solely linked to financial approval and resource allocation. Moreover, it can comprise political support for projects. Examples could include local authority and community support for a particular development initiative, which may be part of a wider regeneration of the local area. In this way, commitment, according to Garbharran et al. (2012), can create a sense of collectivism as opposed to individualism. Consequently, a collaborative environment for projects to be procured within can provide team members with experiences of job satisfaction and has been proven to encourage motivation and promote optimal levels of performance (Challender, 2023). These experiences can provide the right degree of collectivism to prevail on projects for commitment to flourish. Notwithstanding this premise, Garbharran et al. (2012) explained that it is inevitable on any project that changes will occur, and for this reason, adaptability and flexibility are key to refining commitments to take account of these wherever possible.

4. **Communication**
Communication is essential for ensuring that the right levels of leadership, integration of people, collaboration, consultation and decision-making are maintained,

to encourage project success (Garbharran et al., 2012). There should always be a shared vision and expectations for projects. The most successful projects are initiated through a vision workshop where stakeholders discuss, debate and agree on collective goals, objectives, scope and visions for projects. These should be documented for future reference at any stage later in the project life cycle. There can also be thoughts and action plans collated amongst stakeholders as to how to achieve these specific project outcomes. It is usual for project managers to oversee these workshops, but sometimes an independent consultant is appointed, who is not affiliated with the project in any way and therein impartial, and who has experience in facilitating such sessions.

Once the project determinants are agreed and documented, there is a requirement to ensure that they are updated as projects proceed through the various stages. There should be regular progress reports and updates given on meeting the required targets and goals. This is to assure project committees that the defined outcomes are still on course to be delivered against and have not been jeopardised in any way. Project updates and progress reports can be provided by groups or individuals and should be effectively communicated to all stakeholders at regular intervals. Some projects incorporate communication plans as they develop to disseminate information. These can be useful in communicating information to selected groups or to the public through newsletters or websites. The latter is particularly useful where communities have a vested interest in a project and how it is impacting their local area. In addition to dissemination in the case of local authorities and communities, communication should be through direct involvement with these parties at early stages in the project's life. This will encourage political 'buy in' from those in the local area and a feeling that they have been listened to and had an input into future project plans. This is particularly helpful when there is a potentially controversial plan that local residents might otherwise feel could have a damaging and detrimental impact on their livelihoods and the communities that they live within.

9.8 Measuring Performance for Construction Sites

The simplest models to follow in measuring performance are those predicated on the use of KPIs. These are normally associated with attaining the key needs and requirements of an organisation. A performance management system should use these KPIs as the metrics to measure against in terms of levels of attainment and achievement. Salminen (2005) developed a bespoke performance measurement system specifically used for construction sites, as a production unit rather than the organisation as a whole. To support this system, earlier studies conducted by Jonsson (1996), as cited in Salminen (2005) created criteria for assessing construction site performance, and these are included in Figure 9.5.

In considering the factors and criteria in Figure 9.5, it is unusual that the criteria of health and safety and environmental sustainability did not form part of their criteria, which perhaps suggests a gap in the analysis by Jonsson (1996, as cited in Salminen, 2005). Alternatively, these two considerations may not have been considered within the context of the construction industry as prominently as they are in modern times.

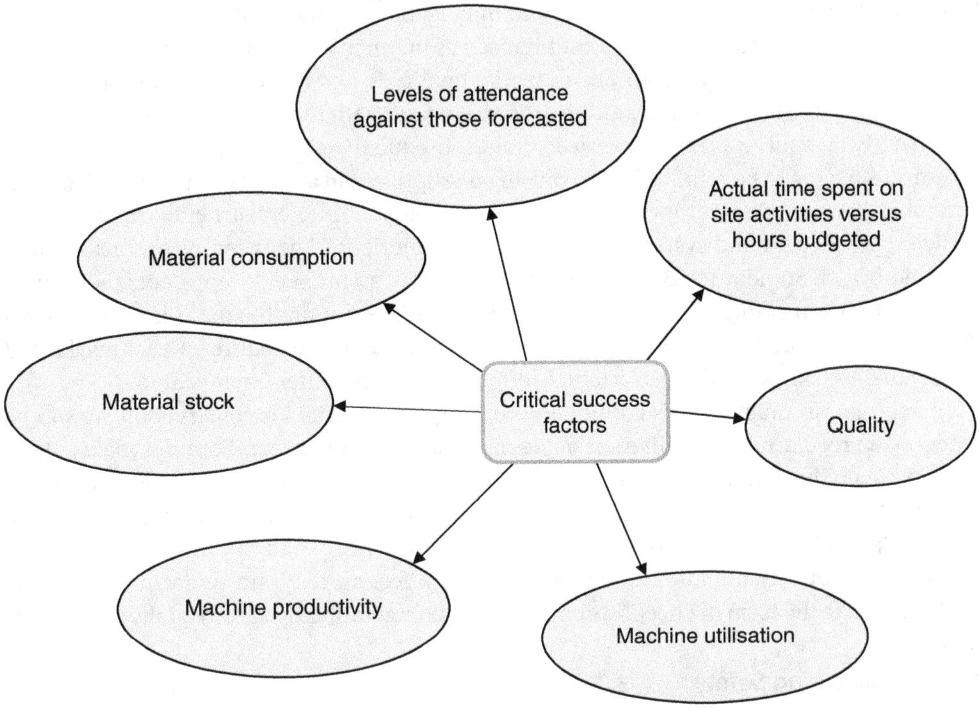

Figure 9.5 Criteria for assessing construction site performance. *Source*: Adapted from Jonsson (1996).

9.8.1 The Dilemma of Measuring Quality on Construction Sites from Different Angles and Perspectives

Sometimes measuring quality can be subjective, owing to the many ways and multitude of perspectives through which assessment can be undertaken. For instance, the extent and degree of quality pertaining to construction work will depend on who is assessing it. Different stakeholders will look at quality from different angles. Clients and end users, for instance, may regard quality pertaining to buildings that function well in occupation, have the least number of faults, snags and breakdowns and require the least amount of maintenance over their life cycles. Conversely, architects and other members of design teams may take a different approach and view quality as an aesthetic design commodity and from the potential to win prestigious construction awards. In contrast, other stakeholders, including local communities and local authorities, may look at quality linked to benefitting the local economy or causing the least amount of disruption to local residents and business owners in the construction phase.

9.8.2 The Importance of Having Good Site Supervision and Quality Assurance on Construction Sites

Site supervision during the construction phases has clearly been regarded as key to achieving the right levels of quality (Salminen, 2005). Any deviations from what could be deemed as acceptable levels of quality for a particular activity can then be identified in a timely manner, through good site supervision and management. This can ensure that any consequential quality issues do not manifest themselves in other works or operations that follow on from them. An example could be to rectify quality issues related to the fireproofing of columns prior to plastering and painting them. If a fault with the fireproofing was not established early through good site supervision and subsequently corrected, then this could lead to finishes having to be removed later, causing disruption and abortive works. It is also commonplace for quality assurance strategies to be integrated into construction management policies, procedures and systems. In the UK, the majority of larger main contractors have adopted British Standard BS5750 with regard to quality assurance. It represents the equivalent in European standards to EN 29000 and in the International Standards Organisation ISO 9000. It is a quality assurance system which sets down procedures, benchmarks and acceptable baselines for the performance of businesses and other organisations.

In addition, in practice, it is commonplace for quality audits for construction sites to be undertaken regularly and at different stages of construction processes. For impartiality, these should ideally be conducted by individuals who are not members of the project team and therein regarded as independent and non-biased. This can then facilitate an open, fair and robust audit where any quality issues can be identified freely and improvement measures agreed, without recrimination or reporting officers feeling they are under duress. These audits can take the form of checklists or questionnaires comprising agreed quality indicators.

9.8.3 Measuring Safety

In the construction industry, it is usual to value health and safety on a par with other KPIs, including productivity and quality. This is clearly the case, as any compromises in health and safety could present obvious risks to construction personnel. Such risks could pose dangers of injuries from unsafe practices or site conditions, and in extreme cases, fatalities. Safety can normally be measured by recording accident frequency against the number of accidents per one million working hours, or alternatively against the number of accidents per 1000 workers. Other factors, such as absenteeism and financial costs attributable to accidents, have also been used in the past. In the UK, the HSE oversees all aspects of health and safety in the workplace, and clearly, this entails construction sites. Where an accident is deemed serious and therein 'reportable', the HSE must be informed immediately. They may conduct investigations to establish the circumstances which gave rise to the incident and assess the level of any health and safety breaches. In some cases, this has led to construction companies being prosecuted for negligence in not maintaining safe operations. This is the case particularly if there are serious injuries or fatalities.

9.9 Summary

There are many different challenges facing the construction industry. These emanate from the bespoke nature of the industry, where teams come together at relatively short notice, undertake a given project and then disband to undertake new appointments. This temporary organisational model presents barriers for long-term relationship building and makes reflective learning very difficult. In addition, some project managers still adopt traditional practices and behaviours which, at times, could be described as adversarial and predicated on win–lose scenarios.

The 'iron triangle' of cost, time and quality has been traditionally referred to as the three success factors for projects. However, what constitutes success on construction projects is not always universally understood, partly due to the complex and dynamic nature of the industry and the subjective nature of success. Accordingly, individuals, groups and organisations will have different ways to judge and measure success, and for this reason, successful outcomes for projects mean different things to different people. In a similar way, there are many different definitions and meanings around what constitutes performance. The measurement of performance can be through 'soft' metrics, such as employee attitudes or 'hard' metrics, which include financial and non-financial metrics. There are many challenges faced by the construction industry in achieving project success and performance. Some of these can be attributable to the highly dynamic environment it operates within, coupled with the changes brought about by new technologies and business methods.

There are many different perspectives and methodologies for measuring project success, and these vary between stakeholder interests. Architects may regard design and aesthetics as the main factors for success, whereas clients and end users may regard operational functionality and delivery of projects on time and budget as more important. Frameworks for evaluating success have been introduced, taking into account the varying nature of projects dependent on their size, value, complexity and use. Criteria for assessment in this regard have included factors such as project time, cost, client satisfaction, conformance with user expectations, functional performance, quality and workmanship and flexibility to users. However, there have been alternative approaches and models related to measuring success from the perspectives of project efficiency, impact on customers, business success and preparing for the future.

KPIs are designed to measure the performance of projects and those organisations undertaking projects and focus on outputs and outcomes which are deemed important and critical. They can be grouped into two distinct areas, around both objective and subjective measures. Objective measures can be more easily measured from tangible data, including construction costs, net present value, time variations, accident rates, environmental impact and unit costs. Conversely, subjective measures could include functionality, quality and satisfaction of end users, clients and project teams. Having KPIs in place can support continuous improvement and establish helpful information and therein create a benchmark for future development programmes and research.

CSFs can be defined as specific actions, areas or elements that teams, businesses or departments should focus on to achieve their strategic objectives. Furthermore, they could be categorised around the 'four COMs' model, developed by Garbharran et al. (2012), which comprises comfort, communication, competence and commitment. The comfort category was concerned with carefully considering and effectively managing the requirements of all project stakeholders. Alternatively, competence is related to the utilisation of modern technologies and knowledge generated from previous projects. The commitment category revolved around senior management investing resources and funding into projects and establishing clear and comprehensive aims and objectives for them. The final category, namely communication, was related to ensuring that the right levels of leadership, shared vision, integration of people, collaboration, consultation and decision-making are maintained, to encourage project success.

The simplest models to follow in measuring performance are those predicated on the use of KPIs. These are normally associated with attaining key needs and requirements of an organisation. Sometimes measuring quality can be subjective, owing to the many ways and multitude of perspectives for which assessments can be undertaken. It is therefore imperative to agree on assessment criteria across all stakeholders. Examples of KPIs on construction sites may include safety, site supervision and quality assurance.

References

Arslan, G. and Kivrak, S. (2008). Critical factors to company success in the construction industry. *World Academy of Science, Engineering and Technology, International Journal of Social, Behavioral, Educational, Economic, Business and Industrial Engineering*, 2(9): 997–1000.

Atkinson, R. (1999). Project management: cost, time and quality, two best guesses and a phenomenon, it's time to accept other success criteria. *International Journal of Project Management*, 17(6): 337–342.

Challender, J. (2022). *Professional Ethics in the Construction and Engineering Industry*. Chichester: Blackwell-Wiley.

Challender, J. (2023). *Organisational Ethics in the Built Environment*. Chichester: Blackwell-Wiley.

Challender, J. and Challender, R. (2024). *Building and Facilities Maintenance Management: An Innovative Operational Approach*. Oxon: Routledge.

Challender, J., Farrell, P., and McDermott, P. (2019). *Building Collaborative Trust in Construction Procurement Strategies*. Chichester: Blackwell-Wiley.

Chan, A. (2001). *Framework for measuring success of construction projects*. CRC for Construction Innovation, Brisbane. Available from: https://eprints.qut.edu.au/216001/ (accessed 23 March 2024).

Egan, J. (1998). *Rethinking Construction. The Report of the Construction Task Force* (pp. 18–20). London: DETR.TSO.

Egan, J. (2002). *Accelerating Change. Rethinking Construction*. London: Strategic Forum for Construction.

Garbharran, H., Govender, J., and Masani, T. (2012). Critical success factors influencing project success in the construction industry. *Acta Structilia*, 19(2): 90–108.

Latham, M. (1994). *Constructing the Team*. London: The Stationery Office.

Nguyen, L.D., Ogunlana, S.O., and Lan, D.T.X. (2004). A study on project success factors in large construction projects in Vietnam. *Engineering, Construction and Architectural Management*, 11(6): 404–413.

Pinto, M.B. and Pinto, J.K. (1991). Determinants of cross-functional cooperation in the project implementation process. *Project Management Journal*, XXII(2): 13–20.

Sadeh, A., Dvir, D., and Shenhar, A. (2000). The role of contract type in the success of R and D defence projects under increasing uncertainty. *Project Management Journal*, 31(3): 14–21.

Salminen, J. (2005). *Measuring performance and determining success factors of construction sites*. Doctoral dissertation, Helsinki University of Technology Construction Economics and Management, Espoo 2005 TKK-RTA-A3.

Sanvido, V., Grobler, F., Pariff, K., Guvents, M., and Coyle, M. (1992). Critical success factors for construction projects. *Journal of Construction Engineering and Management*, 118(1): 94–111.

Shenhar, A.J., Levy, O., and Dvir, D. (1997). Mapping the dimensions of project success. *Project Management Journal*, 28: 5–13.

Vroom, V.H. (1964). *Work and Motivation*. New York: John Wiley & Sons.

10

The Concept of Value Management

10.1 Introduction

To provide context, the chapter will start by looking at the definition and history of value management (VM). VM as a tool for improving project outcomes in the construction industry will then be discussed. This will encompass international perspectives from countries around the world, including Saudi Arabia, which has faced barriers to developing its construction industries in the past. The facilitation of VM will be examined, alongside the methodologies for measuring the output from it. Thereafter, the chapter will discuss the critical success factors for VM implementation, including having a multidisciplinary VM team, a competent VM facilitator, commitment of all stakeholders and effective communication amongst participants. The benefits of VM processes will then be examined from the following different perspectives.

- Encouraging and achieving higher levels of environmental sustainability
- Eliminate unnecessary design work and thus reduce wastage of resources
- Completion of projects within the required budget
- Reducing defects in the final build
- Eliminate unnecessary costs and achieve value for money
- Enhance the value provided for end users
- Encourage environmental sustainability by sourcing materials from local, sustainable ethical sources
- Increase efficiency in construction operations
- Improvements to organisational communications strategies.

The chapter will then move on to discuss critical success factors for value delivery from the perspectives of business management, financial conditions, owner-manager characteristics, quality of work and workmanship, sales and marketing, market selection and use of technology. A model for identifying and tracking value will then be described, which encompasses value aspects from industrial, organisational, societal, professional, project and individual values perspectives. Finally, value frameworks for capturing project success will be discussed around aspects which include internal and external satisfaction and expectations, socioeconomic considerations and environmental performance on projects.

10.2 Definition of Value Management

According to Alsolami (2022), VM can be defined as a multi-disciplinary system which is utilised for the evaluation of tasks associated with particular projects from start to finish, including commissioning. It has also been defined as a tool in delivering multiple objectives through development methodology, innovation and using functional analysis systems.

10.3 The History of Value Management

The first reference recorded with regard to VM dates back as early as 1731, when the academic of his time, namely Daniel Bernoulli, prepared a research paper for the St. Petersburg Imperial Academy of Sciences. The paper presented a case that the value of something should not be purely based on its cost or price, but alternatively on the benefits that it brings. The relationship between value and performance/function underpins all value disciplines. The concept of VM was developed in the 1940s and was introduced within the manufacturing industry in the United States to monitor the performance of projects. This was at a time during World War II when there was a shortage of raw materials, parts and skilled labour. It was adopted by the US Department of Defence in 1954, where it was known as value analysis/engineering. It was at this time heavily scrutinised but was generally considered to offer continuous improvement and flexibility to overcome barriers and challenges, with outstanding results. However, it took until 1996 in the United States when the public policy gave encouragement to companies and government agencies to implement and maintain VM in their processes and events to improve cost effectiveness (Alsolami, 2022).

10.4 Value Management as a Tool for Improving Project Outcomes in the Construction Industry

From the research of Alsolami (2022), a case study was undertaken in Saudi Arabia, where the construction industry accounts for 7.8% of GDP and 16.9% of the workforce, and where it is linked to oil and gas generation and exploration. In this study, there was an articulation that many challenges are faced, including automation and digitalisation. In addition, the increased use of Information and Communication Technology (ICT) can prove daunting for some countries with emerging economies, such as the Middle East, where barriers are faced with the workforce being more conversant, familiar and comfortable with traditional practices and ways of working than in more modern societies. For this reason, Alsolami (2022) found that although Saudi Arabia is committed to high levels of funding to enhance the industry as part of its 2030 Vision Strategy, it requires new methodologies and cultural changes to reach the level of transformation required.

According to Alsolami (2022), the construction industry in Saudi Arabia is plagued by cost overruns, especially in major engineering works such as road construction and realises the need for improved project management practices to overcome this dilemma. This is partly attributable to administration failures, including late payment and insufficient project planning. Accordingly, VM is considered one of the main tools which could be deployed to 'turn the tide' and deliver more successful outcomes for the construction and engineering industries. This would seek to optimise the operational value of projects, looking to maximise all stages of projects from concept to practical completion and in use.

10.5 Facilitating Value Management into Project Management Methodologies

Alsolami (2022) explained that VM can be facilitated by a three-stage process, namely pre-workshop, workshop and post-workshop stages. Furthermore, the workshop stage should be conducted in six individual phases, namely: (i) information analysis, (ii) development, (iii) creation, (iv) evaluation, (v) development and (vi) presentation. All these six phases are required to be undertaken following a very disciplined and structured procedure. These should encapsulate the clients' main priorities and the aims and objectives of the projects. In addition, there should be sufficient stakeholder engagement, and creativity should be encouraged at all times.

10.6 Measuring the Outputs from Value Management Processes

Alsolami (2022) explained that there should be an objective assessment of VM from the perspective of feasibility and effectiveness in order to measure the outputs received from investment into the concept. This is designed to give assurance to clients that their investments are well managed and that they are receiving value for money from introducing VM strategies into their projects. Through evaluation of such VM strategies throughout projects, there is scope to implement adjustments, changes and intervention measures throughout the various processes and stages. Alsolami (2022) explained that this should ensure continuous improvement and vigour to the VM methodology and therein respond to any complications and problems faced. In the United States, data has found that employing VM on projects could generate cost savings of between £1 m and £3 m per year (Alsolami, 2022). Its effectiveness and success could result from many different factors. These include elimination of time wastages, performance improvements, problem-solving ability, reliability and preservation of the project aims and objectives at all times. Furthermore, the VM process should not lose sight of the scope of projects and how they should be performed, as opposed to how projects evolve and are executed. In addition, the process of VM is important for ensuring that the construction sector remains sustainable in the long term, as it is not solely a means of reducing project costs. Moreover, it relies on creating social value in redirecting money elsewhere, possibly into the local community, which can generate other benefits for facilities and infrastructure (Alsolami, 2022).

10.7 Critical Success Factors for Value Management Implementation

There is an established link between organisational learning and the success of companies. Where the performance of organisations is paramount for their success (Challender, 2022). For this reason, it is imperative for organisations to measure performance against pre-determined success factors. In this way, they can then monitor which areas need to be improved where performance has dropped below acceptable levels and take corrective action. This keeps companies competitive and able to generate successful outcomes. Functional analysis has been identified as a critical success factor in attaining VM and leading to cost savings. Having VM workshops can be particularly helpful in identifying other factors which may have a bearing on achieving value for organisations.

The commitment of all stakeholders to VM workshops is regarded as being particularly useful in measuring performance and project outcomes. In this way, professional knowledge and experience of participants' respective disciplines and their willingness to accept changes and innovations can be examined. Notwithstanding this premise, there are many different requirements which will dictate the success and effectiveness of these workshops and these, according to Alsolami (2022), are listed in Figure 10.1.

The main factors of importance for conducting VM initiatives include the selection of teams, collection of data, workshop management, evaluation and decision-making and implementation of the VM proposal.

10.8 Benefits of Value Management Processes

The VM processes can generate many benefits, and these will be identified and explained in this section of the book and are illustrated in Figure 10.2.

10.8.1 Encouraging and Achieving Higher Levels of Environmental Sustainability

Bringing teams together in VE workshops, especially at early stages of design, can have the advantage of assessing which environmental sustainability measures can be feasibly incorporated into projects. This scoping exercise, especially at conceptual design stages, can facilitate open discussion on visions and aspirations for levels of environmental sustainability that the team seeks to achieve. This could include incorporating the use of renewable energy technologies into new buildings to lower carbon generation and life cycle costs through reduced energy usage. Furthermore, the VE workshops could assist in gaining environmental accreditation for projects such as BREEAM and Passivhaus, which need to be factored into projects at very early design stages in order to be effective.

10.8.2 Eliminate Unnecessary Design Work and thus Reduce Wastage of Resources

Project teams can use the VE workshops to better plan buildings according to the visions and aspirations of stakeholders and therein scope and define aims and objectives for projects.

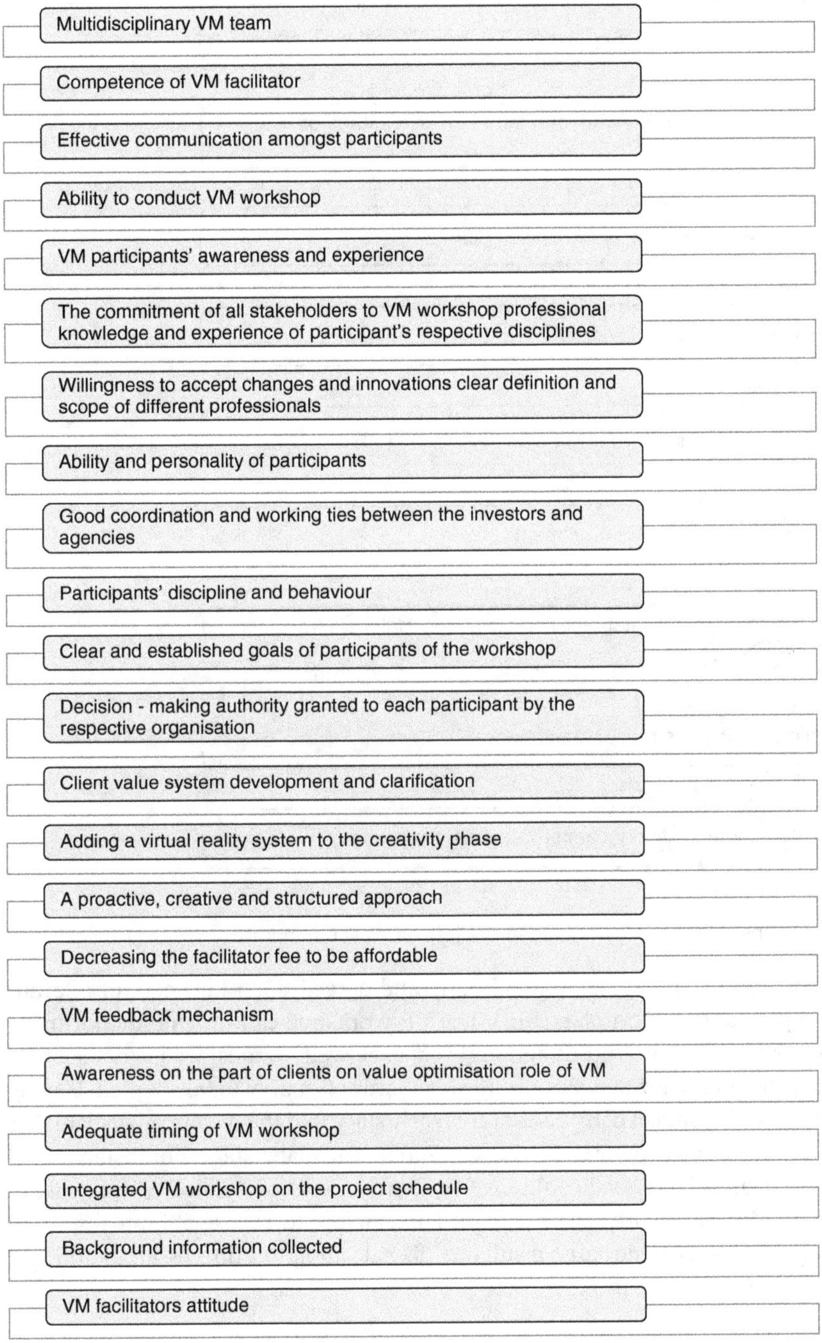

Figure 10.1 Different requirements will dictate the success and effectiveness of value management workshops.

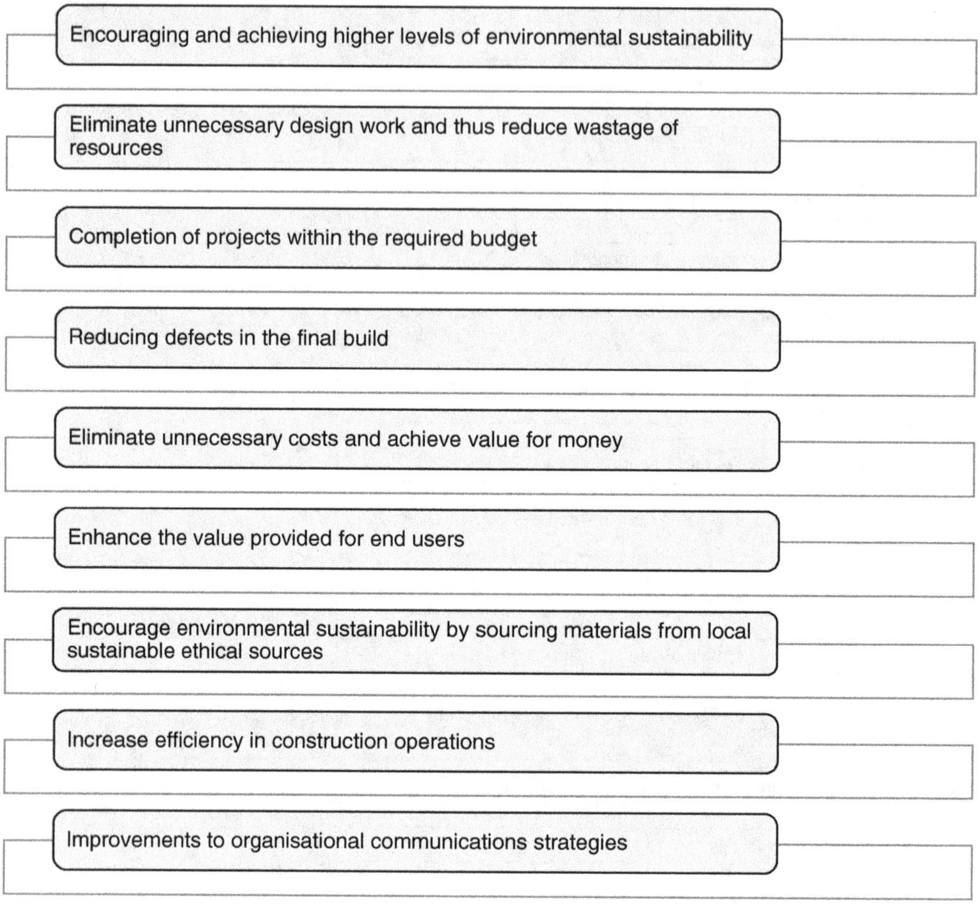

Figure 10.2 Many benefits that value management processes can generate.

It is important for these goals to be captured early and agreed upon between all relevant stakeholders in a project initiation plan (PIP). The VE workshop can provide the forum to debate and agree parameters around design and scope, as well as consider factors such as available budget, programme for delivery and the standards of quality to be adopted. Having an agreement on what the project comprises at this early stage and the resources available for its delivery can save time and money by having to rescope it at a later stage. For instance, setting aspirations at an affordable level could involve abortive work associated with having to undertake expensive redesign work at a later stage. This can lead to a waste of finite resources at the design stage, in terms of design consultants' fees, leading to unnecessary additional costs and possible time delays on projects.

10.8.3 Reducing Defects in the Final Build

The liaison and dialogue between the project team members during VE workshops should enable potential problems and challenges to be explored, leading to problem-solving and better design and construction solutions being established. This can assist in reducing the overall number of snags and defects at the end of projects, as potential issues have been identified earlier on and resolved prior to building completion. Projects which are delivered with fewer building issues at practical completion will ultimately be considered more successful than those that are associated with ongoing issues. This is particularly the case when considered from the perspectives of end users. These stakeholders may regard remedial works to correct defects as unnecessary and disruptive operations that could compromise their use of the building and affect its operation during the works. This can sometimes lead to upsetting the end user occupants through the inevitable post-contract disruption that is caused at these stages.

10.8.4 Completion of Projects Within the Required Budget

VM processes should firstly establish budgets which are workable and achievable for a given project. Budgets should be 'tested' early in the life of projects, normally at the conceptual stage, to ascertain whether they are affordable or not. If they are unaffordable for the expected expectations, then such expectations and aspirations may need to be revisited by stakeholder groups. In this way, it may call for the project to be changed to reduce construction costs, and this may involve simplifying the design or, in some cases, making buildings smaller but more efficient. The nature of buildings, in some cases, may have to be adjusted to compensate for gaps in forecasted project costs and available budgets. Workshops may call for 'Value Engineering' in the way projects are designed to reduce costs. In such cases, applying principles associated with 'Buildability' and other measures to promote best value and increased efficiency. After initial budgets are established and approved, regular tracking and monitoring of budgets by project teams should be undertaken. These are designed to avoid cases where projects start to deviate from established parameters, possibly through overzealous architects and other design consultants, who see them as vanity, potential award-winning schemes on which they can promote their practices. The danger of adopting this approach is that 'scope creep' can be introduced, and projects become unaffordable. If projects appear to be veering in a different direction away from agreed cost plans, then corrective action through the VE workshops should ensure that they are brought back in line as soon as possible. Otherwise, projects can take on a different form than they were initially earmarked to be, which may dictate a reversion back to their original form and nature. Such later changes can clearly result in time and cost implications through unnecessary costs and wastage of resources.

10.8.5 Eliminate Unnecessary Costs and Achieve Value for Money

Achieving value for money can be achieved by good design and ensuring that buildings are performing and achieving those requirements that were established and agreed upon by the stakeholders at early VE workshops. The phrase 'form should follow function' is particularly important to consider for any building. Buildings will not generally be regarded as value for money if they do not deliver functionality for clients and end-users, irrespective of cost.

10.8.6 Enhance the Value Provided for End Users

The previous benefit to eliminate unnecessary costs and achieve value for money will only be truly achieved if end users feel that projects have been delivered to a functional and quality standard aligned with their expectations. For this reason, end users should be included and involved in the VE workshops from the start and be carefully listened to. It has not been unknown for projects to be delivered where there has been very little or sometimes no end-user interface. This is particularly the case with educational projects where complications can arise around teaching and learning accommodation being designed in a way that is not conducive to modern-day pedagogical requirements.

10.8.7 Encourage Environmental Sustainability by Sourcing Materials from Local Sustainable Ethical Sources

Value engineering workshops, especially when they involve interaction and dialogue with subcontractors and the supply chain, can be particularly beneficial where specialist designers, installers, suppliers and manufacturers are involved with specification and design decision-making. In the case of environmental sustainability, these companies are arguably best positioned to advise on the most appropriate and cost-effective materials and components to use for a given application. For example, they may be able to advise on the suitability of a product from the perspective of whether it was ethically sourced or its level of embodied carbon during the production process. Where materials are considered to be environmentally or ethically unsuitable, these specialist companies can advise on alternative sources, materials or components which may offer a greater level of environmental performance. In this way, the environmental credentials of projects can be improved, and this may be important if buildings are earmarked for environmental accreditation, e.g. BREEAM, net zero carbon or Passivhaus.

10.8.8 Increase Efficiency in Construction Operations

Value engineering workshops can encourage individuals to work as teams with the aim of sharing knowledge, expertise and experience and therein achieving much improved project outcomes. Working collaboratively in this way, teams can formulate the best ways and means to construct buildings, taking into consideration site restrictions and parameters and buildability issues. Pooling of different project teams' skills and expertise will sometimes result in different methods of construction being adopted. For a particular project, a modular approach to construction via prefabrication may be appropriate where adverse weather and site conditions are prevalent. This type of construction is generally regarded as allowing quicker installation on site, reducing construction programmes. Its reliance on factory fabrication processes generally can guarantee higher standards of quality control and reduce the associated risk of delays on site from more in-situ related traditional construction technologies.

Other efficiencies brought about by continuous dialogue between project teams may include suitable alternative solutions to site problems. Examples could include cases where adverse ground conditions are experienced and specified groundwork and foundation designs may no longer be appropriate or cost-effective.

10.8.9 Improvements to Organisational Communications Strategies

Teambuilding, collaborative working and partnering are widely regarded as being the conduit that can deliver improved project outcomes through dialogue, relationship building and therein promoting better communication strategies (Challender, 2023). Accordingly, VE workshops, which adopt such collaborative approaches, provide an environment where inter-organisational communications can become particularly effective. Communications between organisations have been linked in the past with relationships that have emerged and developed in collaborative working relationships. Such relationships have provided impetus for professional and ethical practices, actions and decision-making and where values and behaviours and predicated on principles of honesty and openness (Challender, 2022, 2023).

10.8.10 Overarching Benefits of Value Management for Value Management

Considering the above benefits, the construction industry could be greatly improved by the inclusion of VM principles, processes, strategies and policies as an integral part of established project management practices. In this way, its inclusion into what is widely regarded as an industry which is slow to change and adapt to new, improved practices could greatly contribute to awareness and development of construction management. It could achieve this by understanding VM and which critical success factors are most important for delivering successful VM projects and programmes. Accordingly, VM applications in the construction industry can be increased with an overarching aim to improve the quality of construction and value for money. Critical success factors are particularly important in this regard such as providing the framework for strategies around VM to work best and deliver outstanding outcomes for organisations.

10.9 Critical Success Factors for Value Delivery

Arslan and Kivrak (2008) articulated that it is no longer enough to focus on success factors linked predominantly to completing projects on time, cost and quality parameters. Moreover, they concur that there are more factors to be considered in being successful. Some of these relate to management practices and systems. Others relate to the implementation of financial statements and accounting practices. In addition, controls relating to issues such as on-site safety and ongoing training and education of personnel were found to be important for businesses to compete with their competitors in the construction industry. Other factors which have emerged from residential development in the United States were found to have included reputation, good employer-employee relationships, competent and reliable supply chains including sub-contractors, track record of quality workmanship, honesty, client communications and reputation. Furthermore, studies from Germany have concluded that having effective and appropriate risk management, staff development, innovation and research, lean organisational structures and collaborative working relationships through partnering with clients. All these factors have been consolidated and are illustrated in Figure 10.3.

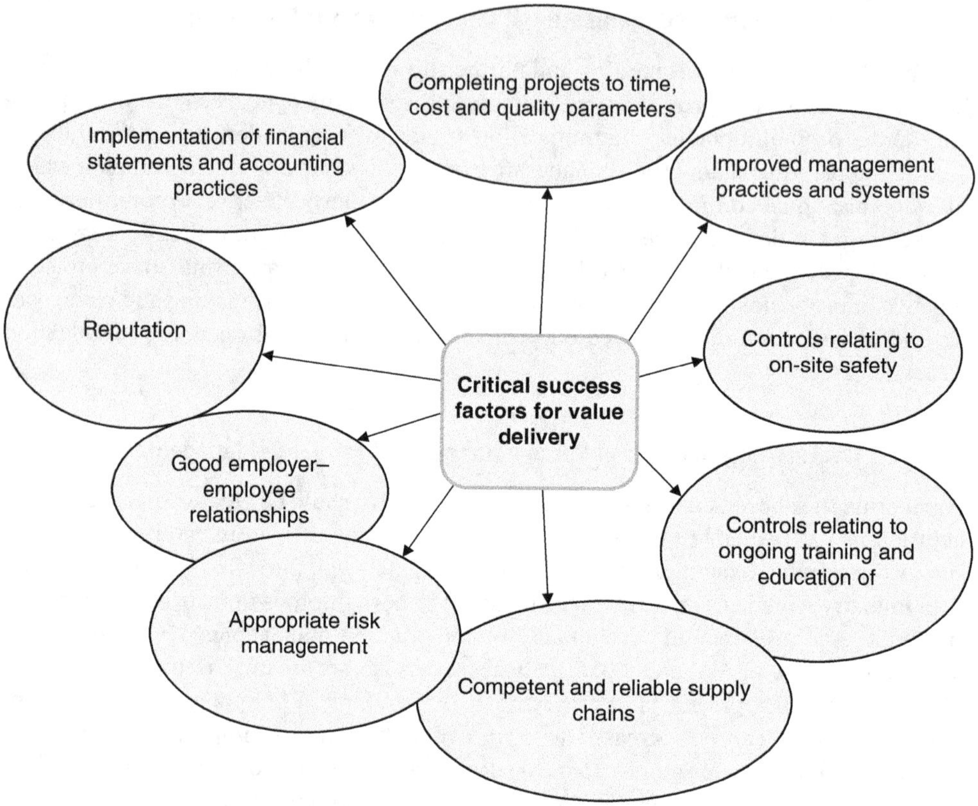

Figure 10.3 Consolidated critical success factors.

The construction industry has traditionally been a risky sector for all those organisations operating within it. The possibilities of projects and businesses failing have always remained high. Accordingly, construction organisations need to consider and focus very intently on those parameters that could have the most beneficial influence on their success. In this regard, Arslan and Kivrak (2008) conducted a research study on 40 construction companies in Turkey to determine the ranking of critical success factors for construction company success. For their research, they utilised seven factors, and the survey participants were required to evaluate the relative importance of each factor to identify which were more important in practice than others. Figure 10.4 illustrates the ranking in order of importance of all seven factors.

The top three factors included business management, financial considerations and owner-manager characteristics, and these ranked over the other four factors. Surprisingly, technology usage was ranked as the lowest for company success, albeit technology was found by the respondents to be an important tool in managing projects.

In addition to these seven factors, Arslan and Kivrak (2008) developed a framework of sub-factors for each, again to identify which were more important in practice than others. The top-ranking factor, namely Business management, was analysed by considering five sub-factors, and the survey findings ranked these in order of importance as illustrated in

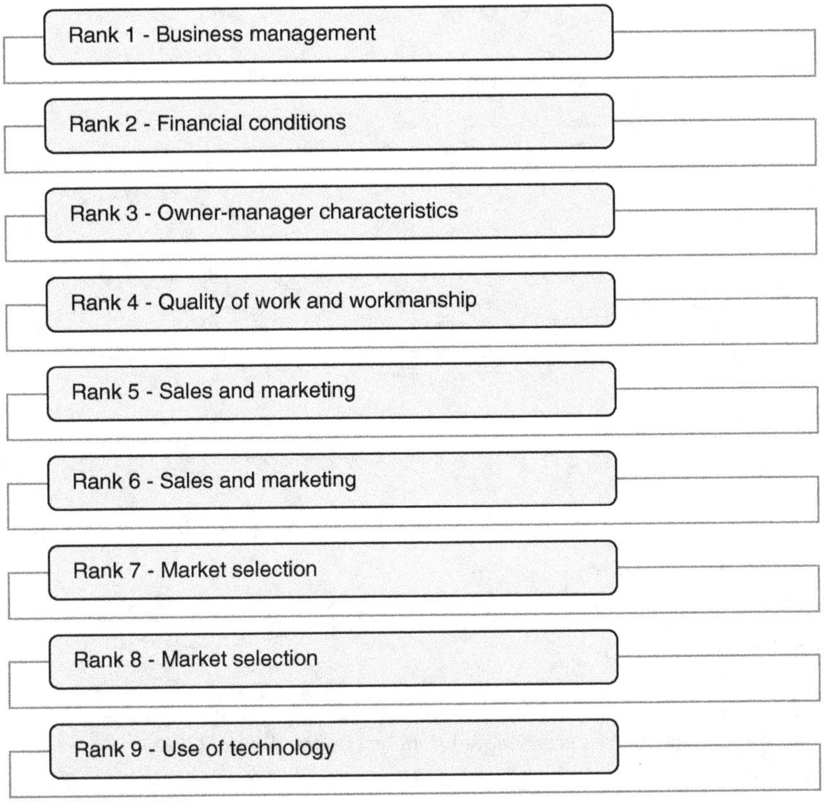

Figure 10.4 Seven critical success factors ranked in order of importance from the study conducted. *Source*: Adapted from Arslan and Kivrak (2008).

Figure 10.5. Organisation and planning came out on top, possibly representing the importance of being able to manage business affairs and construction operations effectively. Conversely, risk management and good record-keeping are regarded as lower down the importance spectrum, which may have reflected that document and risk management are not widely regarded as critical to business success.

The second highest ranking factor, namely Financial Conditions, was then scrutinised from the perspective of sub-factors and illustrated in Figure 10.6. Control and availability of cash flow came out on top in the rankings. This possibly reflected the problems Turkish contractors face with for continuity of their businesses when difficulties arise with late payments, especially synonymous with public sector contracts. Such delays may occur where a lack of funds to finance construction works in the public sector becomes evident.

The third highest ranking factor, namely owner-manager characteristics, was then scrutinised from the perspective of six sub-factors and illustrated in Figure 10.7. Leadership was ranked as the highest sub-factor for company success. This reflected the importance of being able to lead teams of people effectively. Managerial experience of business owners was considered the next most important sub-factor, followed by communication (Figure 10.8).

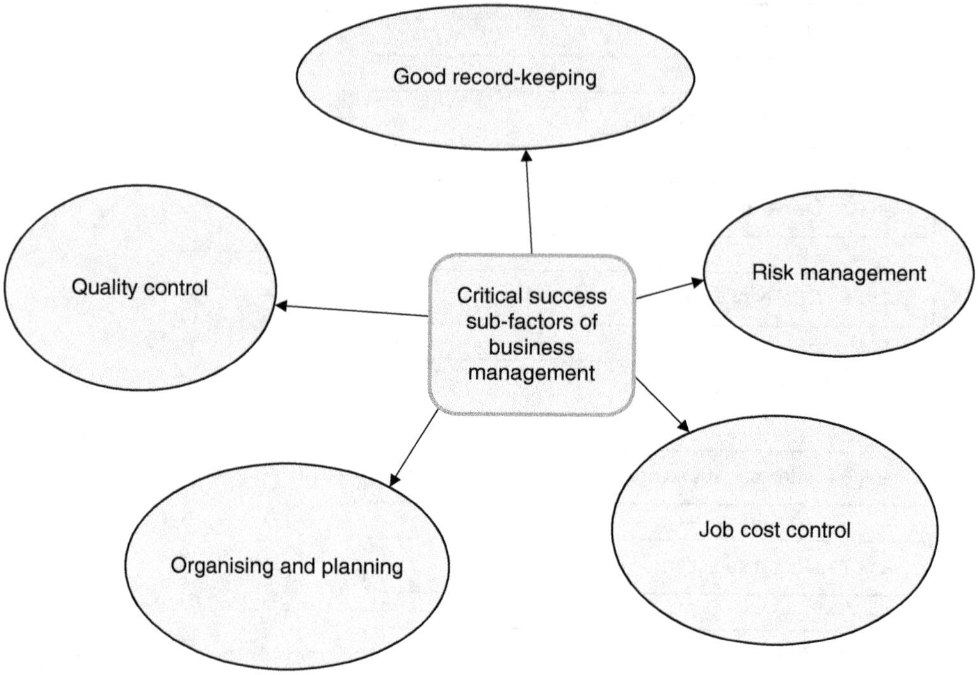

Figure 10.5 Five critical success sub-factors of business management ranked in order of importance from the study conducted. *Source*: Adapted from Arslan and Kivrak (2008).

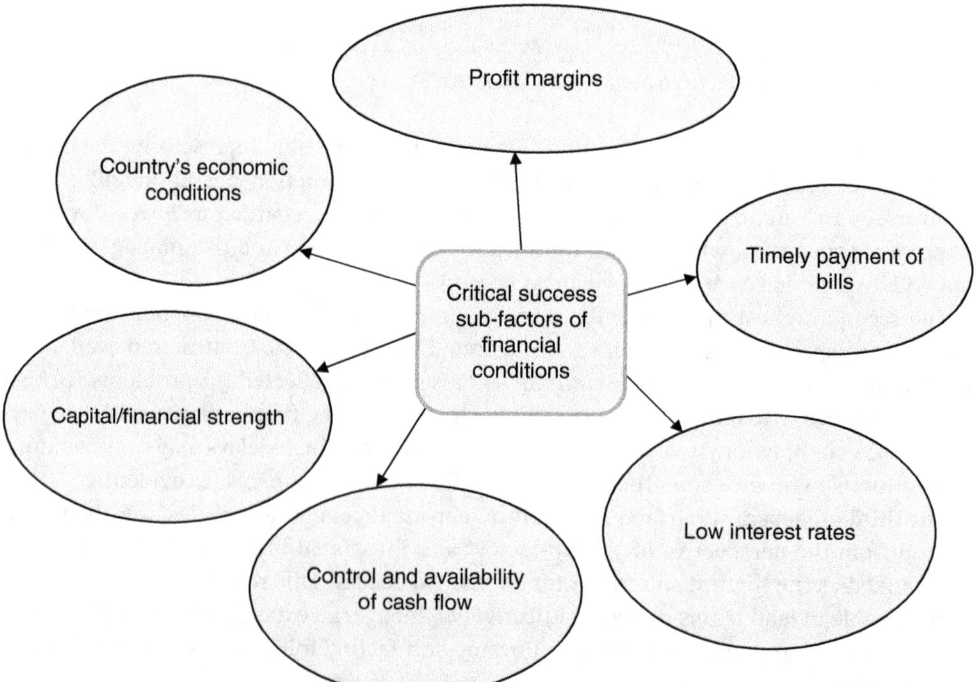

Figure 10.6 Six critical success sub-factors of financial conditions ranked in order of importance from the study conducted. *Source*: Adapted from Arslan and Kivrak (2008).

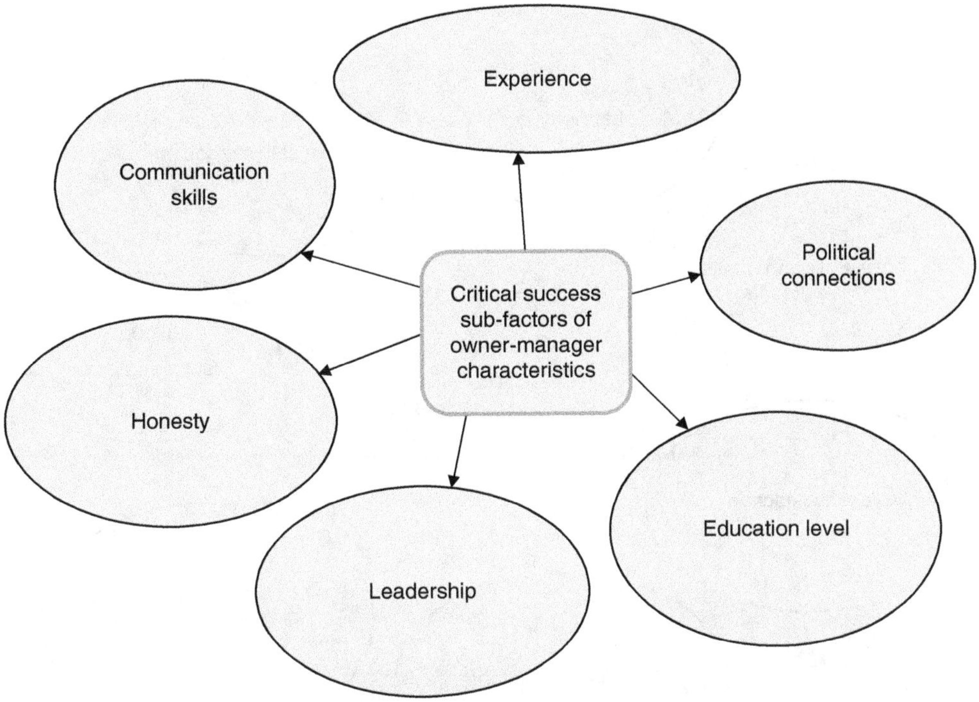

Figure 10.7 Six critical success sub-factors of owner–manager characteristics ranked in order of importance from the study conducted. *Source*: Adapted from Arslan and Kivrak (2008).

When considering the other four factors used in the study, Figures 10.9–10.11 reflect the ranking of their individual sub-factors, ranked in order of importance from the study conducted by Arslan and Kivrak (2008).

For the factor of Quality of Work and Workmanship, rather unsurprisingly, client satisfaction was ranked as the highest sub-factor. This possibly reflected the ability of construction companies, certainly in the case of Turkey, to be able to win future projects. Conversely, having good sub-contractors is ranked lowest for this factor, as possibly considered a norm and a prerequisite in the industry. For the factor of Sales and Marketing, company image was regarded as the most important sub-factor. This again could be related to winning future contracts. Conversely use of modern technologies, such as eMarketing, was ranked as the lowest sub-factor, as respondents did not regard this as having as much effect on the success of businesses in Turkey.

In terms of Market Selection, the company's experience in the market came top of the sub-factor ranking. This could reflect the risks that those businesses that do not possess the relevant experience face in the marketplace, where failures are commonplace. Accordingly, experience is regarded as a critical issue for this reason and a valuable source of giving a business a competitive advantage. When considering the final factor, namely Technology usage, following and adopting new technologies was ranked as the highest sub-factor and having a website was the lowest. This could reflect the operating model in countries such as Turkey, where contracts are won through reputation and previous relations with clients rather than well-presented websites and other technologies.

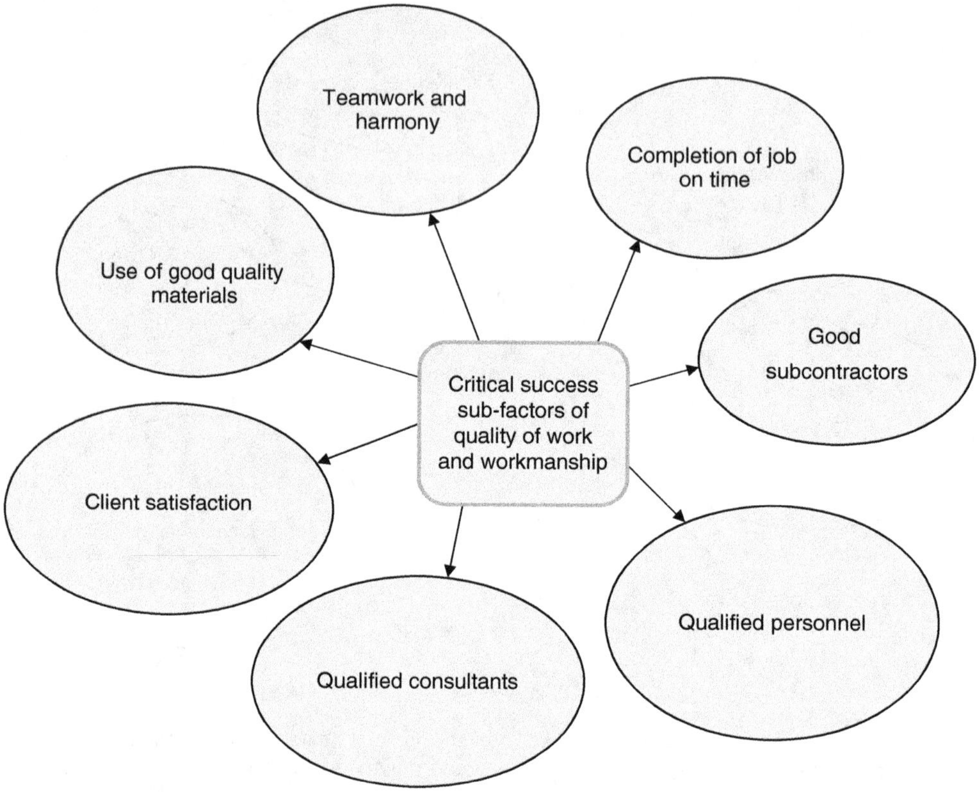

Figure 10.8 Seven critical success sub-factors of quality of work and workmanship.

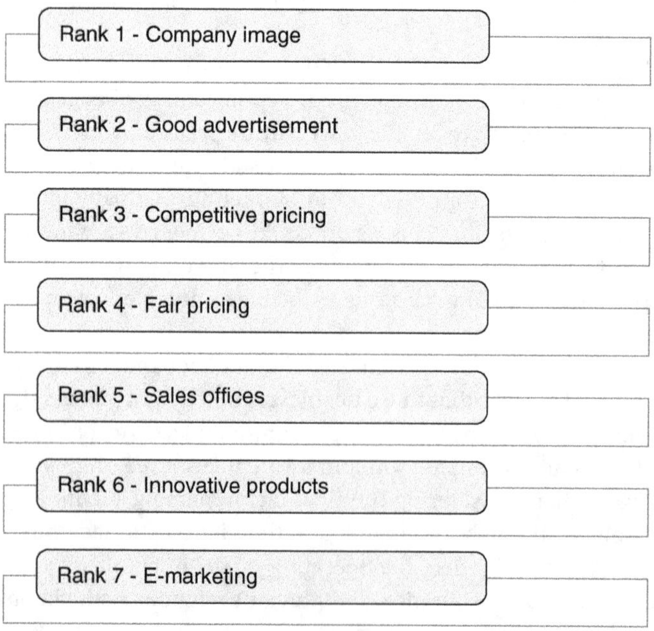

Figure 10.9 Seven critical success sub-factors of Sales and Marketing ranked in order of importance from the study conducted. *Source*: Adapted from Arslan and Kivrak (2008).

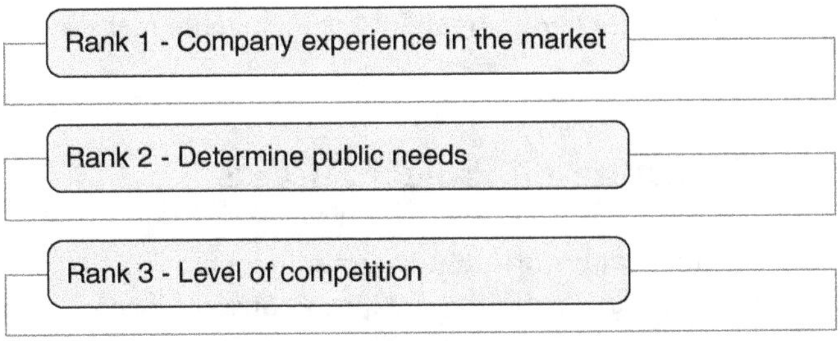

Figure 10.10 Three critical success sub-factors of Market Selection as ranked in order of importance from the study conducted. *Source*: Adapted from Arslan and Kivrak (2008).

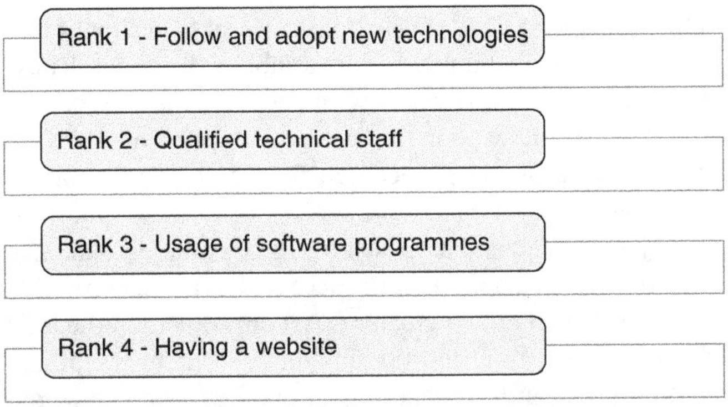

Figure 10.11 Four critical success sub-factors of Market Selection as ranked in order of importance from the study conducted. *Source*: Adapted from Arslan and Kivrak (2008).

10.10 A Model for Identifying and Tracking Value

Khalife and Hamzeh (2020) identified a framework designed to specifically interpret construction projects from six different value aspects, namely industrial, organisational, societal, professional, project and individual values.

Another means for comprehension of the project value concept can be deduced from the analysis of the benefits realised for various stakeholders. This 'benefit analysis' conceptual approach was originally used in the information technology (IT) sector before being transgressed into the construction industry. Traditionally, the benefits from projects have been heavily associated with financial returns received on capital investments, sometimes referred to as 'Returns on Investments' (ROIs). However, in more recent years, there has been a paradigm shift away from solely considering value from the perspective of money, in lieu of other conceivable benefits. These alternative benefits should revolve around stakeholders' requirements and goals and ideally be linked with business strategies. In establishing such needs, there should be a concerted effort at the initial conceptual stages of projects to capture stakeholder requirements and how these can be best achieved through collaboration. Furthermore, it is essential to have an effective means of tracking and measuring value and the achievement of outcomes across the life of projects. This is an important aspect, as previous studies have concentrated solely on creating and capturing value at early stages. In doing so, they have placed little emphasis and failed to explain and address the loss of value during project delivery, possibly brought about by inadequate measuring, monitoring and tracking.

10.11 Value Frameworks for Capturing Project Success

According to Chan (2001), having a value framework that thoroughly and robustly captures such aspects as internal and external satisfaction and expectations, socioeconomic considerations and environmental performance on projects is imperative. Notwithstanding this premise, having robust methodologies for measuring progress against such considerations has become an increasingly important aspect to track performance. However, there is an added complication in practice to undertaking this task. This is related to the fact that over a project lifecycle, and through the various stages from concept to occupation in use, the project participants and stakeholders inevitably and invariably will change. These changes to personnel can present challenges and affect how project value is assessed, as people will see value from different perspectives. For this reason, Khalife and Hamzeh (2020) advocated that due consideration needs to be placed on the relevant stakeholders during each phase and over the lifecycle of projects. They also advocated that the characteristics of each phase should be closely examined and analysed for the creation of value. To illustrate this perspective by way of an example, it is generally accepted in the construction industry that the earlier phases of projects will have a greater effect on value than the later phases. This assertion is predicated on the notion that there is more scope to sculpt and engineer projects around stakeholder goals and expectations at early conceptual stages. In addition, at these earlier stages, Chan (2001) advocated that projects need to be well defined in terms of scope and expected benefits. This supports the views of Arslan and Kivrak (2008) previously articulated in this chapter of the book. Conversely, in later stages, possibly the construction phases of the project, designs have been agreed upon, and any changes to create value from design certainly may be too late to implement. For this reason, value in later phases, including the construction and in use phases, may be lesser but still important from the perspectives of dealing with adversities, challenges and problems and presenting appropriate solutions. In addition, within the IT sector, the 'motivational rationale traceability matrix' developed by Agouridas et al. (2006) has been utilised to align design requirements around stakeholder needs. In a similar way, within the aeronautical industry, a 'Stakeholder Value Network' developed by Sutherland (2009) was developed to comprehend the interactions between various stakeholders and thus capture the flow of common entities around defining, creating and measuring value concepts.

Khalife and Hamzeh (2020) also looked at different ways and means to measure value for different sectors and domains, outside of the construction industry. In the business management sector, the use of a balanced scorecard tool, introduced initially in 1992, has had benefits if integrated within organisational management systems.

10.12 Summary

VM can be defined as a multi-disciplinary system which is utilised for the evaluation of tasks associated with particular projects from start to finish, including commissioning. It has also been characterised as a tool in delivering multiple objectives through development methodology innovation and using functional analysis systems. The concept of VM was developed in the 1940s and was introduced to the manufacturing industry in the United States to monitor the performance of projects.

VM can be facilitated by a three-stage process, namely pre-workshop, workshop and post-workshop stages. The workshop stage should be conducted in six individual phases, namely information analysis, development, creation, evaluation, development and presentation. Projects need to be well defined in terms of scope and expected benefits, and ways of value creation should be agreed upon at early stages, where they can have a greater effect on value than those later phases.

There should be an objective assessment of VM from the perspective of feasibility and effectiveness in order to measure the output received from investment into the concept. This is designed to give assurance to clients that their investments are well managed. Its effectiveness and success could result from many different factors. These include elimination of time wastages, performance improvements, problem-solving ability, reliability and preservation of the project aims and objectives at all times. The VM processes can generate many benefits. These include bringing teams closer together, especially at the workshop stages and scoping projects in terms of measures around environmental sustainability. Other benefits can include eliminating unnecessary design work, reducing wastage of resources and avoiding defects in the final build. VM processes can also help to bring construction costs within the required budgets by assessing whether financial resources and affordability are aligned with projects. In this way, VM processes can eliminate unnecessary costs. Other advantages of VM include increasing efficiency in construction operations and thereby enhancing the value for money for clients. Furthermore, they can lead to improvements to organisational communications strategies and encourage environmental sustainability by sourcing materials from local, sustainable ethical sources.

Value frameworks for encouraging success should robustly capture such aspects as internal and external satisfaction and expectations, socioeconomic considerations and environmental performance on projects. There should be robust methodologies for measuring progress against such considerations, as it has become an increasingly important aspect to track performance.

References

Agouridas, V., Winand, H., McKay, A., and De Pennington, A. (2006). Early alignment of design requirements with stakeholder needs. *Proceedings of the Institution of Mechanical Engineers, Part B: Journal of Engineering Manufacture*, 220(9): 1483–1507.

Alsolami, B.M. (2022). Identifying and assessing critical success factors of value management implementation in Saudi Arabia building construction industry. *Ain Shams Engineering Journal*, 13: 101804.

Arslan, G. and Kivrak, S. (2008). Critical factors to company success in the construction industry. *World Academy of Science, Engineering and Technology, International Journal of Social*, Behavioural, *Educational, Economic, Business and Industrial Engineering*, 2(9): 997–1000.

Challender, J. (2022). *Professional Ethics in the Construction and Engineering Industry*. Chichester: Blackwell-Wiley.

Challender, J. (2023). *Organisational Ethics in the Built Environment*. Chichester: Blackwell-Wiley.

Chan, A. (2001). *Framework for measuring success of construction projects*. CRC for Construction Innovation, Brisbane. Available from: https://eprints.qut.edu.au/216001/ (accessed 23 March 2024).

Khalife, S. and Hamzeh, F. (2020). Measuring project value: a review of current practices and relation to project success. In: I.D. Tommelein and E. Daniel (eds.). *Proceedings of the 28th Annual Conference of the International Group for Lean Construction (IGLC28)*, Berkeley, California, USA. https://doi.org/10.24928/2020/0044.

Sutherland, T.A. (2009). *Stakeholder Value Network Analysis for Space-based Earth Observations*. Massachusetts Institute of Technology.

11

Reflective Practice Around Value Achievement on Projects

The only source of knowledge is experience

Albert Einstein

We do not learn from experience... we learn from reflecting on experience

John Dewey

11.1 Introduction

This chapter will define what is meant by *reflection* and *reflective practice* and help readers understand what constitutes their principles. Thereafter, it will consider the purpose of reflection for project teams and how to reflect on value achievement. In the latter case, it will explore various models of reflection, including Gibbs' Reflective Cycle, Kolb's model of Reflective Practice and the Driscoll Model of Reflection. For each of these models, it will explain the various stages and aspects to help readers differentiate between them. The chapter will then focus on the importance and benefits of reflective practice for projects and articulate suggestions for reflective practitioners. Thereafter, it will consider the implementation of lessons learnt on projects through post-contract review workshops. It will recommend what areas should be covered at these workshops and how lessons from previous projects can make a beneficial contribution to new projects. However, there are limitations, barriers and challenges to the dissemination of knowledge on construction projects through reflection and lessons learnt, and these will be discussed.

Reflective modes will be referred to and the differences and associated benefits of *reflection on action, reflection in action and reflections for action*. Thereafter, reflective processes and levels of intensity applied to projects will be covered and the different levels of intensity around reflection, ranging from *Level 0 – Revisiting* to *Level 3 – Critical Reflection*, and the levels in between. The chapter will then examine the frequency and extent of reflection and challenges faced with projects. This will cover the conditions, and the environment teams are working within and describe the three main categories, namely *Opportunity to Reflect, Ability to Reflect and Motivation to Reflect*. Thereafter, the chapter will transgress to describing the role of a facilitator in the reflection process in conducting workshops and all stages of the reflective processes. Finally, the chapter will explain the dilemma of time availability for reflective processes on projects and how the skills of the facilitator are important for determining the relevance of some issues over others and guiding the whole process.

11.2 What Is Reflection?

There are many different definitions of what we mean by 'Reflection' and 'Reflective Practice'. Self-reflection could be described as the process of introspectively analysing one's own feelings, thoughts and actions to help us gain a greater understanding and insight. In this way, it presents a conscious way to step back and assess one's own decisions, experiences and overall life journey. It is through self-reflection that individuals and teams can identify strengths, weaknesses, patterns and areas for growth or change.

Reflective practice could be defined as 'learning through and from experience towards gaining new insights into oneself and practice' (Finlay, 2008). In this way, it represents a process for reviewing which facilitates links from one event or experience to the next, capturing lessons learn and improving future practice and results.

11.3 The Principles of Reflection

The principles around reflection learning processes relate to looking back at a relevant event and then subsequently analysing it from a personal point of view. In such processes, individuals should then consider their reactions, thoughts and emotions and think about what they can learn from it. Reflective practice, when applied to projects, should examine one's own performance from a personal perspective and combine ability with personal feelings. It should explore how individuals and teams approached an event or issue on a project and consider the strengths and weaknesses of those approaches. Furthermore, there should be an in-depth analysis leading to explanations and a conclusion on what one could do differently to get a better result. It is universally accepted that there are five principles of reflective practice. These revolve around the following areas in Figure 11.1.

11.4 The Purpose of Reflection for Project Teams

The purpose of reflection is for learning from experiences, situations and events of the past and trying to assimilate ways and means through alternative actions that might bring about more favourable outcomes. For instance, if certain aspects of a project were determined to have had less than successful outcomes, one could analyse the circumstances and events leading up to the problems and issues arising. This reflective analysis should consider the factors that decision-making was predicated upon and any other relevant events. Furthermore, according to Hartmann et al. (2023), other aspects should be closely scrutinised from the perspective of any alternative approaches and decision-making that should or did get considered in coming to the agreed course of action. Questions should then be asked and debated if any different courses of action could have had more preferable outcomes.

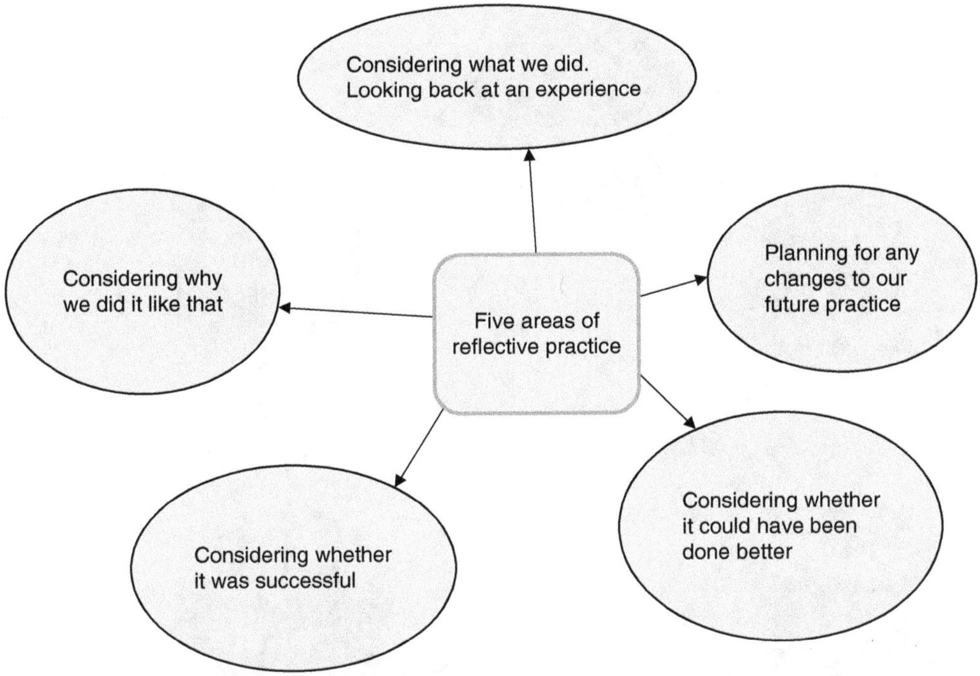

Figure 11.1 Five areas of reflective practice.

If people and organisations can realise the consequences of their decisions and actions, they can hopefully formulate new insights and perspectives which may shape their behaviours and beliefs and allow them to develop as reflective practitioners. Individual reflection provides the medium whereby people can reach their own conclusions on what did and did not go well on a project. In this way, they can be more receptive to change and improve as opposed to feeling criticised if another individual or team were to point out their possible failings. In the latter case, this could be regarded as a negative experience which could lead to defensiveness in behaviours and an unwillingness to accept what could be regarded as blame. In a project-based context, reflective practice can, if practised correctly, enable construction professionals to become improvement agents. In this way, they can deal more effectively with those projects which are particularly challenging, dynamic, uncertain, complex and risky by looking at the issues faced in the past (Hartmann et al., 2023).

11.5 How to Reflect on Value Achievement

There are many models of reflection that can be used to assess value achievement. Some of these will be explained below and are especially important if applied to construction projects.

11.5.1 Gibbs' Reflection Cycle

Arguably, one of the most prolific and well-known models of reflection is Gibbs' Reflective Cycle. The model is based on six distinctive stages which seek to explore the areas of experience, provide a description of such experiences, and feelings around these. Thereafter,

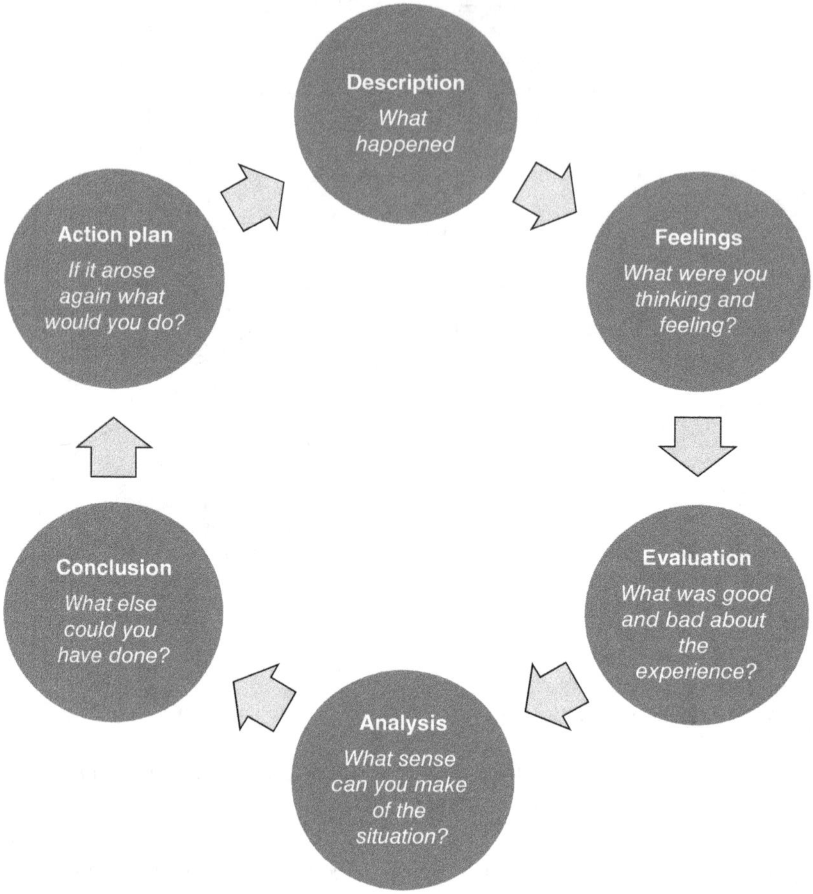

Figure 11.2 The cyclical nature of the Gibbs' model.

other stages include evaluating and analysing those experiences and feelings, reaching conclusions and developing an action plan. This is illustrated in Figure 11.2, showing the cyclical nature of the model, and in Figure 11.3 asks pertinent questions at each stage and therein assists in formulating and writing reflections using Gibbs' model.

Within each of the above stages, several questions can be asked in order to provide information about the incident, event or decision that was made, and these are illustrated in Figure 11.3.

11.5.2 Kolb's Model of Reflective Practice

Kolb's Reflective model has been commonly referred to as 'experiential learning'. This means that the basis of the model is predicated on capturing one's own experiences. Thereafter, such experiences are reviewed, analysed and systematically evaluated, using three stages as illustrated in Figure 11.4. On completion of this cyclical process, any new experiences will form the basis of starting a further cycle.

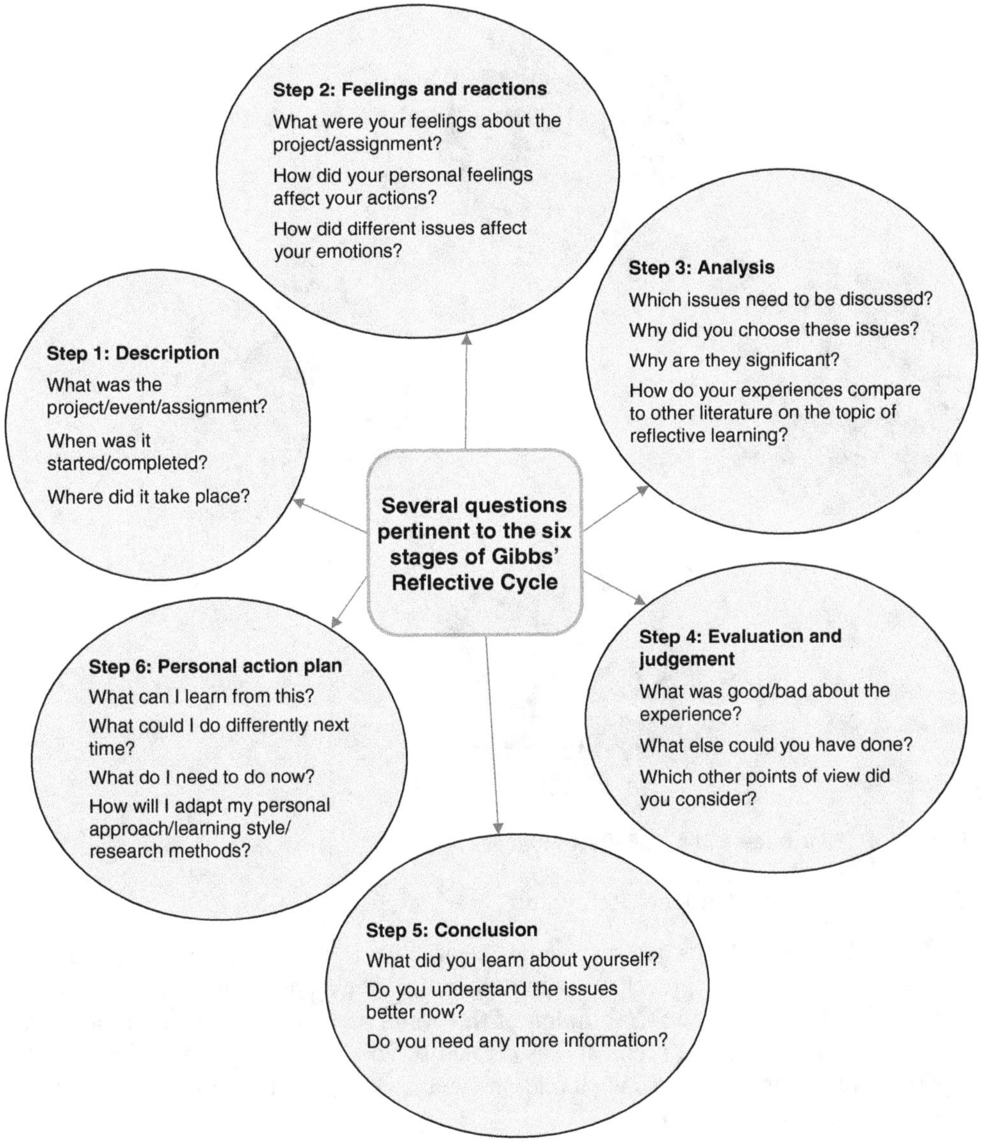

Figure 11.3 Several questions pertinent to the six stages of Gibbs' Reflective Cycle.

11.5.3 The Difference Between the Gibbs' and Kolb's Models

Kolb's model of reflective practice is commonly referred to from the perspective of a model based on experiential learning. Accordingly, it is rooted in learning through experience. Conversely, Gibbs' model is most frequently referred to as an iterative model. This means that it is grounded in learning through repetition rather than experience. This represents an important differentiation between the two models.

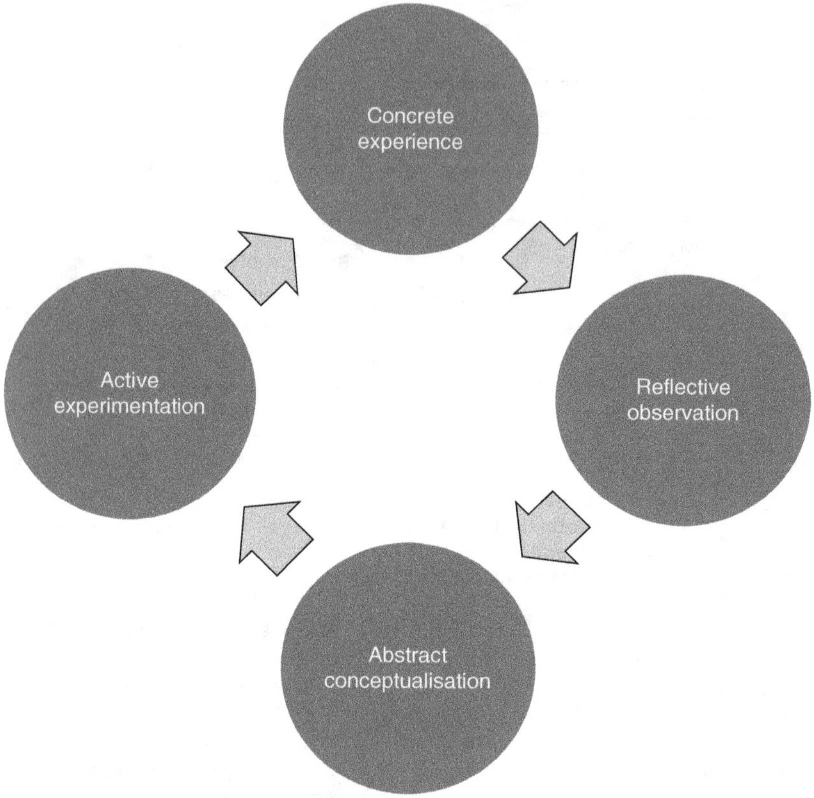

Figure 11.4 Four stages of Kolb's Reflective Model.

11.5.4 The Driscoll Model of Reflection

The Driscoll Model of Reflection represents one of the most well-known and prolific theories that adopts a staged approach to reflective practice. The model includes three elements of reflection, which include a description of the event, an analysis of the event and proposed actions following the event. It was created by John Driscoll in 1994 and asks the important questions of what, so what and now what, as applied to reflective experiences. This is illustrated in Figure 11.5.

11.6 The Importance and Benefits of Reflective Practice for Projects

Reflection is important in order to focus on personal reactions and feelings and therein examine patterns and links in one's thought processes. This is designed to identify how an individual or team approached something and the way they did it. In doing this, the process of reflection should help in understanding how people feel about their experiences and, in this way, recognise their own strengths and weaknesses. In this way, it should develop strategies for the future to overcome weaknesses and build on strengths.

In consideration of the above, the benefits of reflective practice can be that it provides greater awareness of the learning process. In addition, it can create a better understanding

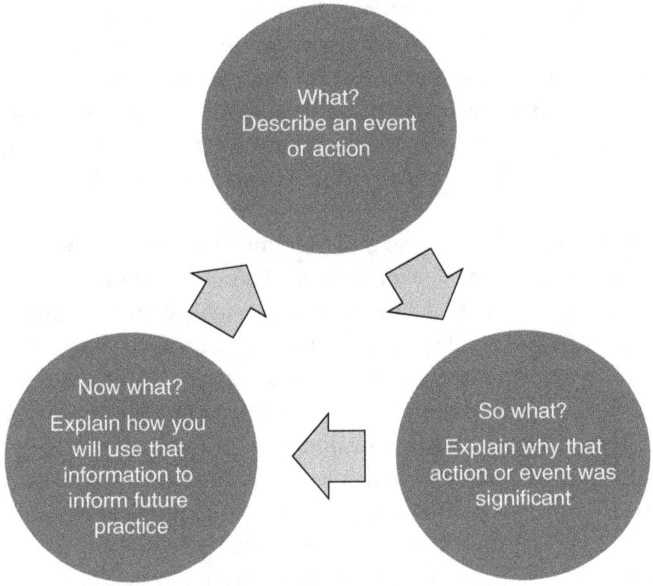

Figure 11.5 The Driscoll Model of Reflection.

of oneself and the issue under consideration. In applying this to the context of the construction industry, this should introduce new perspectives into project management and further encourage best practices and more successful outcomes for projects.

11.7 Suggestions for Reflective Practitioners on Projects

Project management in the construction industry is fast-moving and dynamic in nature. Issues and problems are faced on a daily basis, and responses to these are not always remembered, fully understood or documented. For these reasons, it is important for project participants to keep a learning journal. In this way, they can use it to record their own experiences, reactions and what they learned. These can form the basis of discussions at lessons learnt workshops. Practitioners can, over time, review their journals regularly and ask themselves if their opinions or actions have changed. Furthermore, it should facilitate them sharing their experiences and talking about them. In doing so, this interaction between associates and fellow project team members should make them think more deeply and logically about things.

11.8 Implementation of Lessons Learnt on Projects

Reflection, whilst enabling individuals to prosper and develop, can have far wider benefits if their experiences and lessons learnt can be shared with others. Hartmann et al. (2023) described this as 'collaborative reflection' and they explained that this can be particularly useful if disseminated to other project teams within the same organisation. Where possible, their experiences can become a forum for open discussion amongst teams and lead to a common understanding of project-related problems. In this way, collaborative reflection can be an important mechanism for project learning and encouraging and promoting professional competencies within organisations. Furthermore, professional bodies such as the Royal Institution of Chartered Surveyors and the Chartered Institute of Building have, for many years, tried to create an open form for construction project information to be shared across their members. However, this is not always forthcoming owing to fears at an individual and organisational level of reputational damage by exposing experiences that did not go well.

In terms of capturing lessons learnt from projects, the most common approach in the construction industry is to facilitate post-contract review workshops. These should be attended by all the relevant project participants, including clients, design team consultants, contractors and end users. Sometimes they can be attended by specialist sub-contractors and other stakeholders, where they have had significant involvement. These workshops should present an open forum to freely discuss the project and any relevant events that could be construed as successful or less successful. In both these cases, there should be an examination of the circumstances that led to the event or issue. This should involve open discussion about what could be done the same in the cases of success or differently in the case of any event that was deemed unsuccessful. In the latter case, improvement measures could be delved into a concerted attempt to adopt alternative courses of action on future projects should similar events or issues occur. The workshops should be documented, and the information safely retained for use on future projects. Sometimes, it is good practice on new projects for teams to convene at the earliest stages to look back at documented lessons learnt from previous contracts. This could be at the Project Initiation Meeting, where teams come together for possibly the first time to review the new project from the outset. This is especially important where projects are complex, and similar challenges have been faced in the past. Teams can review those challenges and how they were addressed as foresight for any predicted issues that may arise on their particular project and discuss how best to address these.

11.9 Barriers and Challenges to Dissemination of Knowledge on Projects Through Reflection and Lessons Learnt

Projects have become a way for businesses to organise themselves in many industries, as well as construction. In this way, projects should facilitate organisations to address complications, problems and challenges in a flexible manner. Hartmann et al. (2023) argued that without reflection, an ability to share project-based learning from experiences will not be possible, which could have the effect of reducing the transgression of information to future projects and consequently reduce any scope for industry and organisational learning. They concluded that the bespoke and one-off nature of the construction industry can provide the right platform for new knowledge to be generated but at the same time presents barriers for further continuous improvement.

In consideration of such barriers, the construction projects represent a temporary organisational approach where teams come together for the duration of projects and are then disbanded. Hartmann et al. (2023) explained that this presents a dilemma and limits the extent and degree of information and knowledge dissemination through lessons learnt for use on other projects. In addition, sometimes organisations are not able or unwilling to share any lessons learnt from what could be regarded as failures outside their organisation for confidentiality, privacy and data protection reasons. Others may be reluctant to share experiences where things did not go entirely to plan for risk that their reputation could be damaged. Accordingly, this closed approach of passing knowledge from one project to another could be hampering continual improvement in the construction industry and reducing the overall scope for achieving successful outcomes and achieving best practice. Notwithstanding this premise, reflection has widely been regarded as an appropriate means by which project-based learning can be practised.

In consideration of the above aspects, stimulation of learning-based reflective practices remains a major challenge for the construction industry. In the past, post-project reviews through independent analysis or lessons learned workshops of multiple stakeholders have provided the basis of constructive analysis. However, collation from such reviews can sometimes decouple reflective practice and be devoid of lessons learnt in the way they are conducted and formally documented. This can limit the degree of learning that can be generated for the benefit of future teams undertaking future similar projects. Accordingly, this can cast doubt on the usefulness of the information for learning.

11.10 Reflection Modes

According to Schon (1987), there are two basic reflection modes which are present at different points in time. These are referred to as 'reflection on action' and 'reflection in action'. The former occurs after completing a task where individuals and project teams closely consider and evaluate a task which has been accomplished. They should share their experiences and try to extrapolate conclusions and recommendations for future tasks. This process relies on capturing knowledge and experiences and transposing this information for possible transfer to other projects. As previously referred to in the book, post-contract review meetings, sometimes referred to as lessons learnt workshops, attended by all relevant stakeholders, can be the forum for collecting these lived experiences. Conversely, reflection in action occurs whilst performing tasks and activities rather than after they have been completed. Project team members, rather than reflecting on actions after tasks have been undertaken, will interpret their existing working experiences and situation and reflect on these before adjusting their ways of working to complete ongoing tasks. One could argue that, owing to reflection in action being embedded in ongoing tasks and practices, it cannot be as controllable and repeatable as reflection on action. In addition to the two reflective modes referred to, a third form of reflection was proposed by Schon (1987). This is referred to as 'reflection for action' and involves reconsidering a practice in a manner which is forward-looking to anticipate future situations and issues and to plan a future response.

Reflection for action, according to Hartmann et al. (2023), was found to be the most preferred mechanism for projects. This was explained from the premise that it has the potential to provide deeper insights into the reflective practices whilst facilitating the mechanics to better enable learning. It has the advantage of keeping close contact with the current and ongoing work of project teams. In this sense, it does not rely on retrospective analysis of experiences from previous projects, where recollections may have diminished over time. It has the other benefits of allowing lessons to be applied to the projects at hand and allowing timely address of issues and problems, as opposed to storing up lessons for use on other projects and existing projects suffering as a consequence.

Notwithstanding the above, reflection for action should not be regarded as the 'silver bullet' that can properly ensure that reflective practice is administered adequately and effectively. There is also a reliance on proceeding to later phases and adopting higher intensities of reflection to ensure it is meaningful and useful. This is reliant on time and resource availability, the support given by organisations, the skills of the facilitator and the motivation and willingness of participants to engage in the process. It is recommended that reflection for action is given a prominent place in the procurement of projects and not simply regarded as an afterthought or 'box ticking exercise'. Its part must be centre stage and committed to every level at the earliest stages of projects. Its role does not need to be decoupled from daily working practices, but more so regarded as an enhancement to normal project management business practices.

11.11 Reflective Processes and Levels of Intensity Applied to Projects

According to Hartmann et al. (2023), reflective processes can be categorised by four distinct phases, namely articulating experience, developing an understanding of experience, re-evaluating understanding of experience and drawing reflection outcomes. In addition, research by Fleck and Fitzpatrick (2010) identified that there are different levels of intensity around reflection. In this regard, they identified the lowest level as 'Level 0 - Revisiting', where project individuals simply refer to their own experiences of events or issues without further discussion, dialogue or elaboration. Owing to the limited level of in-depth exploration, this level of reflection is not always useful for project learning. The second lowest level, 'Level 1 – Descriptive Reflection', involves team members describing their experiences but justifying their course of action based on their prior knowledge. This level is also rather limited as it does not seek to explore any alternative explanations, perspectives or courses of action to comprehend their experiences. The next level, namely 'Level 2- Dialogic Reflection', entails team members standing back from their experiences and consciously seeking to analyse these from multiple angles in an attempt to better comprehend and explain such experiences. In this way, they can examine any changes or reorganise existing project practices. The final and highest level, 'Level 3 – Critical Reflection', involves team members questioning assumptions of their own current project practices but extending this to beliefs and values that are taken for granted in the organisation or industry that they work within. This in-depth analysis is sometimes referred to as 'triple-loop learning'.

Fleck and Fitzpatrick (2010) explained that the higher levels of reflective intensity are not always superior to those of a lower level, as this will be dependent on the experiences and issues considered. By way of an example, Level 1 Descriptive Reflection' may be regarded as appropriate to comprehend deviations on projects and reconciling and establishing future corrective actions to address such issues. Conversely, more in-depth reflection by way of Level 3 – Critical Reflection' may prove necessary if such deviations recur frequently and systematically because of organisational systems, procedures, rules and practices, where further in-depth analysis is required.

11.12 Frequency and Extent of Reflection and Challenges Faced on Projects

According to Hartmann et al. (2023), the frequency and extent of reflection practised on projects can be dependent on the conditions and the environment they are working in. Such conditions and environmental factors apply to both the project and team members. In this regard, they identified three main categories, namely 'Opportunity to Reflect', 'Ability to Reflect' and 'Motivation to Reflect'.

The first category, namely 'Opportunity to Reflect', relates to the project environment and conditions being able to provide the most appropriate opportunity for team members to participate in their reflections within. Higher-level intensities of reflection activities can be dependent on having sufficient time and resources made available to reflect. Such reflection should be highly focused and adopt a concerted effort to conduct an in-depth exploration of issues experienced. These issues should also be considered from multiple perspectives, and the analysis should be examined for different explanations. Notwithstanding this premise, finding the time and resources to undertake these activities can sometimes be challenging in busy work environments where pressures of progressing other project tasks may have to take priority.

The other category identified by Hartmann et al. (2023), namely 'Ability to Reflect', relates to the individual skills, abilities, capabilities and knowledge of the team members to conduct the reflective processes. In this sense, the ability to reflect takes into account the fact that individuals need to possess the right degree of abstract thinking and mental capacity to extrapolate themselves from their experience and look inwards. In this way, they can explore the experiences from cause-and-effect perspectives, leading to appropriate assertions and conclusions being drawn. This can entail re-evaluation of experiences and negotiation with other team members, which can call for listening and deploying collaboration and communication skills to make the exercise worthwhile and productive. It also requires an open and honest environment to encourage free speech, where mistakes can be articulated in a genuine and valuable way without threat of recrimination. In this way, understanding around experiences can be properly thought through, considered and appraised to form potential new ways forward for the future. It should be stressed, according to Hartmann et al. (2023), that effective skills do not always come naturally and instinctively, but training from repeated reflective practices can improve them over time. For this reason, they concluded that experience from reflective activities contributes to and improves the ability to reflect.

The final category relating to conditions and environment reflection is classified as 'Motivation to Reflect'. This category captures the motivation and desires of project teams, from an intrinsic and extrinsic perspective, to reflect. Intrinsic challenges may be presented where members may be unwilling or not inclined to engage and participate in reflective practices. This can pose a particular problem as reflection requires the sharing of experiences from different individuals to scrutinise and understand the issues arising, and through discussion, learn from them and formulate more appropriate alternative course of action in the future. Moreover, extrinsic motivational challenges may relate to circumstances where there are external factors which could preclude members from participating in reflective practices. An environment of collaboration is required within any project team to 'open up' and perhaps disclose events pertaining to their experiences, which may be uncomfortable for them, especially if they have not attained good outcomes. To create the right environment to facilitate open disclosures and those conditions to allow free speech amongst team players, confidence and trust are considered essential. According to the research study by Challender et al. (2019), there is a reciprocal correlation between collaboration and trust. Where trust amongst teams is lacking, it would be unlikely that team members would be willing to collaborate and disclose information to their colleagues that could potentially incriminate them, especially if escalated to a higher level of management. For this reason, any fears or anxieties felt by team members could severely compromise any sharing or disclosures they are prepared to divulge if they feel there is a potential threat of retaliation. This can compromise and paternally 'railroad' the whole process. Consequently, creating a collaborative environment of trust and partnership working is crucial in making reflective practices productive and not counterintuitive and pointless.

11.13 Role of a Facilitator in the Reflection Process

In consideration of the above challenges for reflective practice, Hartmann et al. (2023) found that the role of a facilitator for conducting workshops and all stages of the reflective processes is an important facet for successful outcomes on projects. This is particularly the case for more in-depth levels of reflection, such as 'Level 3 – Critical Reflection'. Facilitators who have previous experience in conducting similar sessions can structure workshops to involve all participants. They can also focus attention on extrapolating valuable findings and rich data from the discussions. This facilitating role in the case of reflection-for-action goes beyond simple discussion and evaluation of experiences and actions. Moreover, facilitators require the capacity and capabilities to guide a pathway for the dialogue between participants and offer support to them in being able to open up to reveal their own 'lived' experiences and issues they have faced. This is regarded as being essential to provide the agenda for which other participants can discuss, debate and contribute in a meaningful way. This should generate findings relating to enhancing future practices and adopting different strategies and decision-making, should similar issues and dilemmas occur in the future.

This facilitator role has in the past been undertaken by a representative of the main contractor organisation. In these cases, it is preferable to have employees who are centrally placed within the contracting organisation or involved in other projects and, in this way, are regarded as being completely impartial and unbiased. Alternatively, a completely independent consultant, not affiliated with any one of the project team organisations, may be brought in to oversee the role of facilitator. Participants may feel this independence gives authenticity to the facilitator role as an 'honest broker' and that there are no organisational agendas at play. Accordingly, they may be more comfortable with this approach and, as such, be more prepared to engage with honest, open and constructive dialogue of their experiences.

Facilitators should be able to ask the right questions in providing encouragement to project teams to come forward and share their experiences for the reflection of the wider project team. Such questions should be searching and critical in nature and should delve into the cause and effect of relevant events and activities. In this sense, they should conduct a 'deep dive' into the circumstances leading to decision-making. Furthermore, without providing reflection themselves, reach out to participants to suggest and debate alternative corrective lessons learnt for the future. Sometimes the facilitator is a merged role for an individual who currently is performing a different role within an organisation. If this is the case, due consideration should be given to whether they have the right capabilities and enough time and resources to fulfil the role properly. Training and education may become appropriate if they do not possess the right skills, and where possible, their workload in other areas should be reduced to compensate for their expenditure of time as a facilitator.

11.14 Dilemma of Time Availability for Reflective Processes on Projects

Hartmann et al. (2023) concurred that time is a prevalent factor in providing opportunities for reflection processes but also in allowing enough time and resources to conduct them in a sufficient and meaningful way. In this sense, generally speaking, the more reflection phases coupled with the higher reflection intensity, the more time is required.

The skills of the facilitator could play an important part in progressing the phases of discussion and reflection at a pace which adheres to time availability, given that time is a finite and precious resource, and time spent reflecting must compete with other countless priorities. This requires the skills of the facilitator in sensing the relevance of some issues over others, guiding the whole process and knowing when each issue and stage has been sufficiently covered to be able to move on to the others. In this regard, there may be instances presented where there is a tendency to rush the processes in a concerted attempt to reach premature findings and conclusions. Furthermore, in some cases, workshops could become interrupted through time pressures, potentially stopping discussions and forcing the parties to find outcomes that cannot be fully understood or deemed credible. This could have the effect of carrying over potentially new lessons learn and changes to working practices that are either not fully substantiated, authenticated or justified or alternatively having reached the wrong outcomes and conclusions. In the latter case, especially refining practices and changing organisational way of working, this could be particularly damaging. Such damage could be predicated by reflecting on the wrong outcomes of workshops. Where this occurs, it can have severely detrimental consequences and lead to further issues and negative experiences recurring.

11.15 Summary

There are many different definitions of what we mean by 'Reflection' and 'Reflective Practice'. Self-reflection could be described as the process of introspectively analysing one's own feelings, thoughts and actions to help us gain a greater understanding and insight. Alternatively, it could be defined as learning from experience and gaining new insights into oneself and practice. The principles around reflection learning processes relate to looking back at a relevant event and then subsequently analysing it from a personal point of view, considering reactions, thoughts and emotions. When applied to projects, there should be an in-depth analysis leading to explanations and conclusions on what individuals and teams could do differently to get better results in future.

The purpose of reflection is for learning from experiences, situations and events of the past and trying to assimilate ways and means through alternative actions that might bring about more favourable outcomes. This scrutiny from the perspective of any alternative approaches should benefit future decision-making. In this way, individuals and teams can deal more effectively with those projects which are particularly challenging, dynamic, uncertain, complex and risky by looking at the issues faced in the past.

There are many models of reflection that can be used to assess value achievement. The *Gibbs' Reflective Cycle* is a model based on six distinctive stages, namely *Description, Feelings and Reactions, Analysis, Evaluation and Judgement, Conclusion and Personal Action Plan.* The model seeks to explore the areas of experience and provides a description of such experiences and feelings around these. *Kolb's Model of Reflective Practice* is predicated on capturing one's own experiences, commonly referred to as *experiential learning*. The *Driscoll Model of Reflection* includes three elements of reflection, namely a description of the event, an analysis of the event and proposed actions following the event. The benefits of reflective practice are that it can introduce new perspectives into project management and further encourage best practices and more successful outcomes for projects.

Reflection, whilst enabling individuals to prosper and develop, can have far wider benefits if their experiences and lessons learnt can be shared with others, sometimes referred to as *collaborative reflection.* Where possible, their experiences can become a forum for open discussion amongst teams and lead to a common understanding of project-related problems. In terms of capturing lessons learnt from projects, the most common approach in the construction industry is to facilitate post-contract review workshops. These should be attended by all the relevant project participants, including clients, design team consultants, contractors and end users. They should involve discussions about what could be done the same in the cases of successes or differently in the case of any event that was deemed unsuccessful. Notwithstanding the previously defined benefits, there are barriers and challenges to the dissemination of knowledge on projects through reflection and lessons learnt. These may emanate from people's reluctance to share experiences where things did not go entirely to plan, for risk that their reputations could be damaged.

Reflection on action occurs after completing a task, where individuals and project teams closely consider and evaluate a task which has been accomplished, whereas *reflection in action occurs* whilst performing tasks and activities. Conversely, *reflection for action* involves reconsidering practice in a manner which is forward-looking to anticipate future situations and issues and to plan a future response. Reflective processes can be categorised by four distinct phases, which include articulating experience, developing an understanding of experience, re-evaluating understanding of experience and drawing reflection outcomes. In addition, there are different levels of intensity around reflection, namely *Level 0 – Revisiting, Level 1: Descriptive Reflection, Level 2: Dialogic Reflection and Level 3: Critical Reflection.*

The frequency and extent of reflection practised on projects can be dependent on the conditions and the environment they are working. There are three main categories, namely *Opportunity to Reflect, Ability to Reflect and Motivation to Reflect.*

The role of a facilitator for conducting workshops and all stages of the reflective processes is an important facet for successful outcomes on projects, given some of the aforementioned barriers and challenges. They can provide the capacity and capabilities to guide a pathway for the dialogue between participants and offer support to them in being able to open up to reveal their own 'lived' experiences and issues they have faced. Facilitators should be able to ask the right questions in providing encouragement to project teams to come forward and share their experiences for the reflection of the wider project team. The skills of the facilitator could play an important part in progressing the phases of discussion and reflection at a pace which adheres to time availability, given that time is a finite and precious resource, and time spent reflecting must compete with other countless priorities.

References

Challender, J., Farrell, P., and McDermott, P. (2019). *Building Collaborative Trust in Construction Procurement Strategies*. Chichester: Blackwell-Wiley.

Finlay, L. (2008). Reflecting on reflective practice. *Practice-based Professional Learning Paper 52*. The Open University.

Fleck, R. and Fitzpatrick, G. (2010). Reflecting on reflection: framing a design landscape. In: *Proceedings of the 22nd Conference of the Computer-human Interaction Special Interest Group of Australia on Computer-human Interaction* (pp. 216–223).

Hartmann, A., Vinke-de Kruijf, J., and van Weesep, R. (2023). Asking the right questions: the role of reflection for learning in and between projects. International Journal of Project *Management*, 41(5): 102494.

Schon, D.A. (1987). *Educating the Reflective Practitioner*. San Francisco CA: Jossey-Bass.

12

Case Study of Reflective Practice Applied to the Construction Industry

12.1 Introduction

This chapter represents an example as a case study of reflective practice of the book's joint author, Professor Jason Challender. It will explain the various models and stages of reflection, referred to in the previous chapter, and how these were practically applied to a project by Professor Challender. The project under *reflective investigation* relates to a gap in current academic provision for the construction industry.

Through reflective practice, needs analysis and research, Professor Challender identified a gap in current academic provision as far as construction teaching and learning is concerned and with particular emphasis on undertaking building inspections and providing survey reports. This chapter is intended to provide a workable example of how the gap was identified through the background and reflections of Professor Challender, and the subsequent stages that led to addressing the problem. Accordingly, it will give context to the author and his personal reflections which established the gap in current academic provision. Furthermore, with reference to academic sources gained from a literature review, it will detail the various stages of investigating the problem and implementing the project, with reflective practice as its driver. The 'project' as such was not a construction project, but for Professor Challender to create a 'surveyors handbook' to address the problem associated with the gap in academic provision. This chapter will include the following stages of this project as illustrated as follows.

- Personal reflection stage
- Research, supporting evidence and justification for the project
- Planning and devising the project,
- Mapping against Lifelong Learning UK (LLUK) standards
- Implementing the project
- Feedback, evaluation and personal/professional Development

Finally, the chapter will provide a conclusion through a personal reflective statement from Professor Challender, following completion of the project. This outlines the success of the project and how the surveyors' handbook may be able to address the gap in the academic provision and be used and benefit construction management and surveying students in the future.

12.2 Context to the Background of Professor Challender

Professor Challender with over 35 years' service as a senior director in many different organisations where he has played an important advocate to the construction industry. He established through reflective practice and through his own experience and research, that a gap in academic provision exists. As both a practitioner and visiting professor, he sought to explore why teaching and learning have not been adapted to meet the needs of the construction professions and particularly building surveying students over the years. Furthermore, he developed an innovative project that he piloted on the Extended Diploma Construction Studies Course at Stockport College in 2013 which he sought to enhance teaching and learning for the benefit of himself, his students and other tutors. Following its integration and implementation, he then invited feedback and evaluation on whether it had, in fact, made a difference and enhanced vocational training for construction students. In this chapter, there will be an explanation of the personal and professional development that he has gained from the project and how this has benefitted him and his students as both an author and visiting professor.

12.3 Personal Reflection

The first stage of considering how to progress a project and develop his professional practice as a practitioner, author and visiting professor was for him to undertake reflective practice. Ferraro explained:

> *The primary benefit of reflective practice for teachers is a deeper understanding of their own teaching style and ultimately, greater effectiveness as a teacher. Other specific benefits noted in current literature include the validation of a teacher's ideals, beneficial challenges to tradition, the recognition of teaching as artistry, and respect for diversity in applying theory to classroom practice.*

Taking account of Ferraro's approach above, his reflective practice focused and looked back at his own education whilst at College and University to examine whether students on leaving education are being fully prepared for employment and thereby competent to undertake their chosen vocations.

His own personal experience, as a building surveying graduate in 1989, was that on leaving university, he was not equipped with the basic awareness of how to actually undertake inspections and surveys of buildings. This was despite completing a BSc Building Surveying degree over a period of four years full time. This was quite surprising for his employers, a large multidisciplinary firm of chartered surveyors, who had to coach and train him through the building surveying process. He was, however, at this early stage of his career required to undertake many different types of building surveys including structural surveys, schedule of dilapidations and schedules of condition. Accordingly, being able to competently and professionally undertake this type of work was expected of him in professional practice. This presented somewhat of a dilemma for him at this point in his career. He was expected to carry out work which was regarded as 'routine' for a graduate building surveyor of his calibre, but without any prior knowledge or experience in how to undertake it. Personal reflection identified that there was a critical gap in his abilities to carry out his role with the practice, and this called for him requesting support and help from his employers at this time. Notwithstanding this deficiency, the expertise and training he received in professional practice enabled him to undertake building surveys through training from senior site management colleagues who had experienced similar dilemmas at the same stages of their careers.

Burgum and Bridge (1997) explained that a happening during one's professional practice which has the potential to yield meaning if analysed properly could be commonly regarded as a 'critical incident'. They explained that an incident is critical in the sense that its analysis yields meaning, trends, motives, etc. This is reiterated in Moon (1999) when she stated that:

> *Critical incident analysis is a specialised form of 'revisiting' of previous learning or experience in which individuals or groups unpick a significant event or an incident and reflect upon it to come to understand what they have learned from it.*

Furthermore, Boud et. al. (1985) advocated a three-stage approach to recording and describing a critical incident through recollecting the experience, attending to feelings and re-evaluating the experience. Using this staged approach to consider the 'critical incident' led Professor Challender to ask the question as to whether there is a gap in provision as far as teaching and learning of students on construction and surveying related courses is concerned with particular emphasis on preparing technical building surveying reports. Simple reflective practice from his experiences, those of his past and present colleagues in industry and his recent interface with construction studies lecturers and students led him to the suggestion that there is currently a large divide between classroom teaching/academic qualifications and the basic skills required in professional practice.

Professor Challender, during the early stages of his teacher training and completion of a Post Graduate Diploma in Education (PGDE), was required to be observed by peers whilst carrying out some of his teaching sessions. Feedback from colleagues, coupled with his own personal reflections, reinforced his beliefs that not enough practical surveying-based skills are being developed in students and too much emphasis is placed on construction theory in the classroom. His reflections also led him to the theory that a closer link between academia and professional practice will motivate learners as his own class, although enthusiastic about future vocations, found it hard to see the relevance of construction and surveying classroom teaching. To reinforce this theory, there is research to show that when traditional teaching and learning methods are employed, students lack motivation and fail to progress beyond rote learning (Lavelle, 1992). Professor Challender's reflective thoughts on these findings were that students should be stimulated during the learning process in order to optimise their attention, performance and learning, and that his proposed project must address this aspect accordingly.

This retrospective reflection on concrete experience was supported by Kelly (1997), following Kolb's learning cycle theory, where concrete experience is evaluated and considered at the 'reflective observation' stage. This then allows conclusions to be drawn of what we have learned as part of the 'abstract conceptualisation' stage, and then trying out our new learning through 'active experimentation'. This will be referred to in more detail in Section 12.7 of this chapter.

There are other theories that adopt a staged approach to reflective practice including John Driscoll's (three elements of reflection) model as referred to in the last chapter of the book. This model, to reiterate, advocated a description of the event, an analysis of the event and proposed actions following the event. Peters (1991) also described a similar process called DATA that consists of four steps of reflection, namely:

- Describe the problem, task or incident that represents some key aspect of practice that the practitioner desires to change
- Analyse the information to establish patterns, meanings, insights, clues and trends.
- Theorise to draw conclusions, inferences, judgements or evaluative statements.
- Identify appropriate action to change behaviour by way of a new and modified approach.

Using these step processes led to a 'reflective enquiry' by Professor Challender and triggered several questions that he would seek to unravel as part of his project. These included:

- Where is the evidence that justifies the idea that theory and practice need to come together in construction teaching?
- Why is there a gap in provision?
- What improvements could be made in teaching provision to bridge the gap?

12.4 Research, Supporting Evidence and Justification

The next stage was to investigate whether there was any evidence to support and give credibility to Professor Challender's beliefs through his research and answer some of his reflective questions.

He undertook research via the internet, academic books and research papers, construction journals, the Construction Skills Council, the Chartered Institute of Building (CIOB) and the Royal Institution of Chartered Surveyors (RICS).

His initial research was quite alarming and revealed that:

- 8% of employers represented within the Construction Skills Sector Skills Council have faced hard-to-fill vacancies largely due to low numbers of applicants with the required skills.
- 18% of building contractors within the Construction Skills SSC have traditionally suffered from internal skill gaps, with staff being regarded as less than proficient (National Employers' Skills Survey, 2012).
- Only 25% of students enrolling on CIOB partnership courses in Construction Management go on to become Chartered Construction Professionals (Mair, 2005).

The research also identified that the gap in provision, as far as teaching Construction Management is concerned, seems to have existed for many years. Professor Mole's *Mind the Gap* report (1997) led to the RICS Education Reforms of 2001, which sought to address the gap in practical based construction course content. This clearly did not go far enough as a study was carried out by Hoxley and Wilkinson in 2006 to review the RICS imposed changes. Their findings were that employers believed that 'graduates still lack sufficient knowledge' and lecturers seem to be 'teaching what they want to teach rather than what students need' (Hoxley and Wilkinson, 2006).

Wilkinson and Hoxley (2005) concluded that concerns exist about the adequacy of provision of 'core content' within academic construction courses. They suggest that this could have been brought about through pressures related to the limited extent of practical teaching that can be included within the curriculum. They identified that this has largely resulted from the increased pressure to maintain high staff/student ratios and increase module sizes due to under-funding in the UK higher education sector. Their research paper also advised that there has been a genuine interest expressed by all employers in engaging more with universities to address the gap in practical based provision, but concerns were raised that universities were non-receptive to approaches from industry in this regard. In addition, the RICS were aware that 'students cite a lack of basic skills sets and are sometimes unable to make the adjustment from an accredited course to practice' but whilst welcoming diversity and not being prescriptive in the subject/modules of a course '...it expects a course to meet standards of quality, evident with the development of construction skills' (RICS, 2010).

As part of carrying out further research, Professor Challender liaised and consulted with various individuals whom he had access to in his position at this time as Project Director of Stockport College, including professional practitioners and course leaders. There was a resounding belief from these individuals that there is still a long way to go, as far as improvements in teaching construction students are concerned, to meet the skills that are required by industry.

12.5 Planning and Devising the Project

When planning the project, Professor Challender considered a number of factors that would have to be considered. As stated by Reece and Walker (2007):

> *'The decision-making process you go through in designing for teaching and learning is complex and may be influenced by such aspects as students' behaviour, course aims, course content, time available for learning, teaching strategies to be used, assessment techniques required and resources available.'*

In order to address the gap in construction provision, he considered it would be useful to develop an innovative and meaningful teaching and learning resource, which would then hopefully enhance the existing teaching programme for the benefit of learners. Whilst aiming to ensure that 'objectives should describe what the student is expecting to be able to do in order to demonstrate learning' as defined by Walklin (1990), he sought to develop and fulfil the following SMART objectives that I had devised for the project:

- Create a student survey Handbook to assist learners in preparing and building survey reports.
- Enable students to become familiar with the process of conducting building surveys, in both a professional and diligent way, to better prepare them for professional practice.

- Close the current gap between education/training traditionally gained by students at colleges and universities and those skills required by the construction and surveying industries.

A project plan proposal form was prepared, which mapped out the aims, objectives, resources and timescales.

The project itself was essentially focused on the development of a simple and concise handbook to assist students in preparing building inspections and survey reports, and thereby improve their effectiveness in conducting their work in practice. The handbook incorporates concise headings as a guide and template to give structure and to assist students. It would also serve to make students more aware of the methodical and systematic approach to carrying out building surveys, and the pathological approach in the identification and diagnosis of building defects. This was intended and designed to assist them in providing a holistic pathway through the building surveying process via checklists. This was to ensure that elements and components of buildings are inspected and not simply missed from their reports. It was also devised to steer students through a logical and methodical sequence for conducting their surveys and in the process gain knowledge of the various risks and how these are best managed. Its generic format was also designed to enable students to input information about their particular site in the required sections to produce site specific details for their particular report.

Furthermore, as students are generally not familiar with undertaking surveys of this kind, I included questions in each section to prompt responses from learners. Examples of this included description of the site, types of accommodation, different elements and condition of buildings and characteristics and presence of defects.

The students that Professor Challender taught were generally from diverse and deprived backgrounds and generally lacked any work-related experience. Accordingly, for reasons of inclusivity, as well as the SMART objectives, he designed this project with a variety of 'soft' learning outcomes, such as improved confidence, freedom from social isolation and improvement in interpersonal skills. These intangible results of attending learning programmes have their own intrinsic value but can also 'assist progress towards harder outcomes' (Dewson et al., 1997 in V. McGivney, 2002). Self-confidence, in particular, has been seen as crucial to making progress in learning; it promotes 'more ambitious goal setting and more risk-taking' (Eraut, 1994 in V. McGivney, 2002).

In planning and devising the project, Professor Challender had extensive dialogue with his PGDE mentor and course supervisor at Stockport College. This, he believed, added credibility and gave him direction for channelling his reflections into the creation of my project. Atherton (2005) endorsed this approach as follows:

> '...*It can be argued that 'real' reflective practice needs another as mentor or professional supervisor, who can ask appropriate questions to ensure that the reflection goes somewhere and does not get bogged down in self-justification, self-indulgence or self-pity.*'

Professor Challender had agreed and planned with the course leader of the Building Surveying Module (as part of the Extended Diploma Construction Course) at Stockport College that the project could be piloted with students using the handbook as a template to assist them in understanding and addressing the process of carrying out surveys and inspections. It was planned that the handbook would enable students to undertake their own building surveys and, through simulated role-playing activities to equip them for the industry. Professor Challender developed a questionnaire that will be completed by participating students and lecturers to obtain feedback as to the usefulness of the surveyors' handbook.

The basis of this handbook, as a reference tool in carrying out building surveys, was planned to give all the necessary information that the student requires to enable them to undertake 'real life' site specific issues at a later stage when they graduate into industry. Furthermore, it was structured to give students a basic familiarity and understanding of the risks associated with working within a potentially challenging and dangerous environment. This would then hopefully benefit them in their studies and better prepare and equip them for their chosen vocations in life.

In designing the project for these Extended Diploma students, Professor Challender sought to refer to Bloom's Taxonomy theory (1956) relating to the cognitive domain, for assistance in using appropriate language for the level of the students. Using Bloom's theory, he closely considered developing the lower and appropriate higher order skills of the students when preparing the handbook. As these students piloting the project were Extended Diploma Level 3, he adopted the application level of cognitive complexity using terminology such as interpret, apply, illustrate and demonstrate to assess student learning. This then categorised the level of learning in terms of the expected ceiling for the project.

12.6 Mapping Against Lifelong Learning UK (LLUK) Standards

At the time of undertaking the project in 2013, standards were set down by Lifelong Learning UK (LLUK), which is the body for professional development for those people working in further education. The LLUK document on the new professional standards for teachers sought to make clear what is expected of teachers in the sector in terms of skills, knowledge and attributes. The underlying ethos was expressed by Robson (2006) as follows:

> *'Teachers in the lifelong learning sector value all learners individually and equally. They are committed to lifelong learning and professional development and strive for continuous improvement through reflective practice. The key purpose of the teacher is to create effective and stimulating opportunities for learning through high quality teaching that enables the development and progression of all learners.'*

Professor Challender outlined the professional knowledge and practice that he had met as part of this project assignment and mapped out where his project values have been utilised. It is especially important for LLUK standards to be followed in teaching, whilst creating an effective stimulating environment to promote learning. The overarching professional standards for those who teach in the lifelong learning sector cover six domains, namely:

- Professional Values and Practice
- Learning and Teaching
- Specialist Learning and Teaching

- Planning for Learning
- Assessment for Learning
- Access and Progression

These standards provide a consistent framework across the sector with the overall aim of increasing the standard and quality of teaching. They embrace important issues such as equality and diversity, and always maintain an inclusive and motivating learning environment for the benefit of the student.

In consideration of the LLUK standards, it was important for Professor Challender to consider these standards when devising his project in order to meet as many of the professional values as possible. Accordingly, his project was devised to add value and enhancement to current construction and surveying studies teaching by utilising his own professional skills and creating a creative and innovative learning resource designed to motivate students. Reflective practice, needs analysis, research, collaboration with others and creating and implementing an innovative tool for learning have all demonstrated my commitment to the standards.

12.7 Implementing the Project

As far as piloting the project was concerned, Professor Challender prepared some guidance notes and conducted a session with the Extended Diploma students to fully explain the methodology behind the handbook and how to make best use of it. He carried out a similar exercise with Stockport College's course leader and a construction professional from industry, both of whom had considerable knowledge of conducting building surveys, to obtain feedback from a different perspective.

In order to enthuse and motivate learners as part of the sessions when piloting the handbook, Professor Challender provided a safe and comfortable learning environment, encouraged team working, gave praise for student input, outlined future opportunities (future rewards) and tried wherever possible to instil positive thinking. He also gave regular positive feedback to encourage an upward cycle of learning as far as motivation, success, reinforcement and self-belief are concerned (learning engine theory). These factors followed from Maslow's (1954) theory on the pyramid of intrinsic and extrinsic things that motivate.

Keeping up motivation is a challenge for both strong and weak learners, as is finding the right sort of material to suit everyone in such a diverse group. As Minton (2005) found:

> *'One of the problems of putting students of all abilities into the same classes... is the tendency for the achievements of everyone in the group to find a level that fails to realise the potential of the more able among them.'*

According to the Retention and Learning Pyramid theory of E. Dale (1972), students are least likely to retain information when solely relying on conventional lecturing. They are most likely to retain knowledge when teaching others, followed by practising by doing, discussion groups and demonstrations. For this reason, when implementing the project,

Professor Challender tried to encourage students to work in groups and have discussions where appropriate. He also encouraged those students who had already completed tasks related to the handbook to 'Buddy' with those who were clearly struggling. He believed this to be a good teaching strategy as it enabled him to keep everyone up to speed and avoid a minority falling behind, whilst giving the other students the opportunity to teach their peers, which, according to Dale, would reap great benefits for them in retaining knowledge. This is a method of ensuring knowledge is acquired, advocated by Minton (2005), who found that the brain 'selects' only about 10% of the information presented to it. Accordingly Minton (2005) suggested that getting students to explain a particular point to others provokes '...alerting of the brain..., and formulation into terms, words and symbols, that are the students' own.'

12.8 Feedback, Evaluation and Personal/ Professional Development

12.8.1 Feedback

Following this implementation process, individuals including the Extended Diploma students, the course leader and the construction professional, were required to complete questionnaires that Professor Challender had specifically designed for the purpose of obtaining their feedback on the handbook. Written e-mailed feedback responses from the course leader and professional practitioner were also received.

Feedback is a very important part of the assessment and learning process for reflective practices. According to race, Brown et al. (2005), it ought to be developmental, and 'should open doors, not close them'. For this reason, the questions on the questionnaires were carefully worded to enable him to evaluate against a list of predetermined criteria that he had devised as measurable outcomes. These were in turn related to his three main objectives and which would seek to assess the usefulness of the handbook.

Professor Challender also conducted a session with students to review their completed feedback on the questionnaires and to discuss how useful they found the handbook. Students do not, however, like to complete questionnaires which resulted in generally poor participation. There is also the difficulty in collating and evaluating all the individual feedback information that is generated, and how to react and address all comments. Feedback from learners could, with hindsight, have been more effectively generated via a student representative who would collate all students' views and concerns and present this as group feedback. This would then have overcome the problem of receiving potentially conflicting individual feedback from all members of the class. Students may also have been more comfortable in sharing their views and presenting new ideas to a student representative than directly with the tutor, and this, therefore, would have had the advantage of increasing participation.

Positive feedback from all participants included the following:

- The handbook was clear and unambiguous.
- The structured checklist and text box format of the handbook allowed site building survey information to be easily compiled for a given site.

- It promoted a logical and methodical approach to conducting building inspections and preparing professional survey reports.
- It clearly differentiated sections for quick reference.
- The concise information contained in the handbook would assist in the process of carrying out site surveys in a uniform, consistent and professional way.
- As a guide, the surveyor's handbook would ensure that all relevant information is included and no parts are left out.
- It contained excellent cues and guidance to aid completion.

In addition, the advantage from an educational perspective is that learners reported that without the guide steering them through the process item by item, they would have found it more difficult in preparing the full written building survey, which was tasked to them through an assignment brief.

The professional practitioner and course leader felt the guide was probably more conducive to teaching and learning than as a tool for industry in its existing simplistic form. These individuals did, however, feel that the handbook could potentially save a lot of building surveyors' valuable time on site in collating information more swiftly, rather than relying on conventional note-taking if developed further.

Some development points were, however, forthcoming from their feedback and they included:

- Provision of drawings and including site layout plans, to better identify the location and extent of defects.
- The text boxes provided were not adequate to fit the amount of information required.
- Certain questions would have benefited from being reworded for clarity.
- If the handbook were to be developed for professional practice, it would need to contain more information on the condition of some elements of buildings e.g. roof and walls.
- The handbook could have used graphical aids as a teaching resource.

12.8.2 Evaluation

Having reflected on the comments above, Professor Challender found that he would need to address the points through the following actions when compiling a similar handbook next time around:

- Include drawings and possibly aerial photographs as appendices to the main document.
- Remedy the text box space restriction by including expandable boxes.
- Address the issue of clarity through better use of clearer terminology in the document.
- Improve the handbook as a teaching and learning resource by including coloured diagrams and pictures wherever possible. This will also serve to enhance its presentation and visual interest.

In addition to the above, Professor Challender believed the handbook could also have been more inclusive as a learning resource and designed to anticipate the needs of dyslexic learners. Using pastel coloured paper and adopting the 'Comic Sans' font as recommended by Reece and Walker (2007) could have addressed this issue, and this, from his reflections, is something that he would incorporate in the future.

As far as the feedback questionnaires were concerned, Professor Challender also felt that he could have created a more quantitative format by introducing degrees of responses. For instance, rather than a response of yes or no, ask the participants to indicate whether they strongly agree, agree, disagree or strongly disagree with a particular statement. This would have developed and facilitated a more detailed pattern and percentage of responses, and this will be something he will seek to introduce in future questionnaires of this kind.

Hopefully, the different stages of Professor Challender's project, where he had reflected on previous experience, decided to improve and set out ideas for improvement, will benefit him and his students for the future. Accordingly, he intends to carry out those ideas on future projects, demonstrating an adoption of David Kolb's four stage model of learning, known commonly as the reflective cycle theory. This learning cycle, as explained by Kelly (1997), is entered at the concrete experience stage, where the learner is involved in an experience. Kolb argued that to learn from this, we all move to a phase where that experience is evaluated. This is the reflective observation stage. From these, conclusions are drawn about what we have learned from the experience, and we devise a new set of rules or ideas and are ready to try them out. Kolb also advocated that we move from the abstract conceptualisation to active experimentation, where we try out new learning. This may not happen immediately, but trying out can create a new experience and the cycle starts again. Schon (1983) and Gibbs (1988) developed similar theories to Kolb, and these also focused on revolving ideas of reflective practice from one's own experiences.

As far as Professor Challender's desired SMART objectives for the project were concerned, he believed the first two were successfully achieved to his satisfaction, as the handbook clearly assisted learners through each stage of the surveying process and familiarised them with the format of building inspection reports. As far as the third objective was concerned, he believed that the handbook went some of the way to bridging the current gap between education/training gained by students at college and university and those skills required by industry. However, he concluded through his own reflective practice that this is only the tip of the iceberg and believed that much more development of practical skills and learning aids needs to be introduced into construction-related curriculum, generally to better train and prepare students for employment in the future.

By encouraging creativity, as far as developing an innovative learning resource is concerned, Professor Challender felt that his project in devising a surveyor handbook as a teaching and learning guide had the potential to bring about the following economic and social benefits:

- Make students more marketable.
- Offer them more potential for pursuing future vocations.
- Allow them to secure employment in the construction industry.

Furthermore, he concluded that there also needs to be more consultation between the CIOB and the RICS, as representative institutions of the profession, and academic establishments, to advise on improvements as far as accredited course content is concerned. In his role as an RICS Assessor at the time of completing the project, this was something that he wanted to pursue in the future as part of an agenda for change.

12.8.3 Summary of Personal and Professional Development

From his reflections after action, the main benefits for Professor Challender as a specialist teacher at the time and visiting professor now in completing this project are summarised below:

- He believes that he has developed reflective practitioner skills using steps and principles outlined by educational theorists such as Kolb, Schon and Gibbs. This has enabled him to undertake abstract contextualisation from experience with a view to creating detailed actions as to how to improve. This, he believes, has developed his classroom practice and will be a process that he hopes to become more proficient and familiar with throughout his whole teaching career as a visiting professor.
- In creating and implementing the project, it has enabled him to teach more effectively as a specialist lecturer in construction. He also believes that it has given him the confidence and ability to use the techniques and experience that he has developed to create similar innovative and creative learning resources in the future.
- He believes that using the surveyor's handbook will improve the way he conducts his future lectures as a visiting professor in his specialist subject area by conveying information in a manner that can be easily understood by construction and surveying students.
- He believes that the handbook will enable him to be more aware of current technological developments in the construction industry, which will allow him to keep students updated on current developments.

In light of the above, Professor Challender, on completion of the project, prepared a personal reflective statement, as a conclusion, to encapsulate the lessons learnt from feedback and evaluation, and this is contained below.

Personal reflective statement from Professor Challender, following the completion of the project
In general terms, I believe the handbook to be a useful, simple and innovative teaching and learning aid for worthy inclusion within the curriculum. I believe that I have identified a deficiency in my specialist subject area, and this is an opinion that is shared by the Course Leader who plans to incorporate it as part of the Building Surveying module. It offers students a useful tool as relative beginners to the task of carrying out building surveys, possibly for the first time, and hopefully narrows the gap between traditional classroom teaching and those skills required by professional practice. I know from my own experience that tasks of this kind can be bewildering when undertaking them for the first time and I believe that the handbook will hopefully make the process more comprehensive for students in the future. I can foresee its future development opportunities for professional practice as well as for education as it presents a more effective and efficient means of collating building information on site. It has the potential to become integrated by way of palm held computers for instance to electronically input details on site rather than rely on hard copies. This technology could be networked and downloaded on to a software system that could more effectively prepare a survey report automatically. This would make

the whole process so much more efficient and greatly reduce the resources and time currently expended by surveying staff in preparing the completed document. This is particularly relevant in the current economic climate, in construction and surveying especially, where companies are seeking alternative ways to reduce costs and become more competitive. Furthermore, it has the potential to be adapted for different levels of learning and compiled as part of a CD resource pack or online.

As far as my own post evaluation reflections are concerned, I have learned much from the process of developing this project and this has been well documented throughout this report. As part on my ongoing continual professional development (CPD) I will strive to seek out and identify further opportunities and potential projects, utilising the knowledge and experience I have gained, in pursuit of improvements to future teaching and learning in my own specialist area.

References

Atherton, J.S. (2005). *Learning and teaching: reflection and reflective practice*. Available from: http://www.learningandteaching.info/learning/reflect.htp (accessed 24 February 2011).

Bloom, B.S. (ed.) (1956). *Taxonomy of Educational Objectives, the Classification of Educational Goals- Handbook 1: Cognitive Domain*. New York: Mc Kay.

Boud, D. et al. (1985). *Reflection: Turning Experience into Learning*, 3rd ed. Buckingham: Open University.

Brown, S., Smith, B., and Race, P. (2005). *500 Tips on Assessment*. London: Routledge Falmer.

Burgum, M and Bridge, C. (1997). Using critical incidents in professional education to develop skills of reflection and critical thinking. In: R. Pospisil and L. Willcoxson (eds.). *Learning Through Teaching. Proceedings of the 6th Annual Teaching Learning Forum*. Murdoch University, February 1997. Perth: Murdoch University. online (accessed 10 January 2011).

Dale, E. (1972). *The Retention and Learning Pyramid*. London: Falmer.

Eraut, M. (1994). *Developing Professional Knowledge and Competence*. London: Falmer.

Gibbs, G. (1988). *Learning by doing: a guide to teaching and learning methods*. London: FEU. Available from: http://www.glos.ac.uk/el/philg/gdn/gibbs/index.htm (accessed 21 January 2012).

Hoxley, M. and Wilkinson, S. (2006). RICS education reform in the construction industry: the employers' view. In: E. Sivyer (ed.). *COBRA 2006, Proceedings of the Annual Research Conference of the Royal Institution of Chartered Surveyors*. University College London, 7–8 September 2006. London: RICS.

Kelly, C. (1997). *David Kolb, the theory of experiential learning and ESL*. Available from: http://www.aitech.ac.jp/-iteslj/Articles/Kelly-Experiential/ (accessed 24 February 2011).

Lavelle, D. (1992). *Promoting student learning and motivation via a systematic introduction to the use of IT*. In: B. Sloan and D. Schofield (eds.). *Proceedings of the Conference: Computer Based Training in Property and Construction*. University of Salford, September 1992.

Mair, W. (2005). *Built Environment Group, APC Competency Review, Construction Faculty*. London: RICS.

Maslow, A. (1954). *Hierarchy of Needs Motivational Model*. New York: Harper and Row.

McGivney, V. (2002). *A Question of Value: Achievement and Progression in Adult Learning*. Leicester: NIACE.

Minton, D. (2005). *Teaching Skills in Further Education*, 3rd ed. London: Thomson Learning.

Mole, T. (1997). *Mind the Gap- Report on the Rationale and Expectation of the Construction Industry in the Education and Training of Chartered Surveyors*. London: RICS.

Moon, J.A. (1999). *Reflection in Learning and Professional Development: Theory and Practice*. London: Kogan Page.

Peters, J. (1991). *Strategies for Reflective Practice. ITT Module: RRP Section1: Introducing Reflective Practice*. Bolton Institute of Higher Education Department.

Reece, I. and Walker, S. (2007). *Teaching, Training and Learning; A Practical Guide Incorporating the New Professional Standards*, 6th ed. Sunderland: Business Education Publishers.

RICS (2010). *Keep Learning- A Framework for Construction Professionals*. London: RICS.

Robson, J. (2006). *Teacher Professionalism in Further and Higher Education*. Abingdon: Routledge.

Schon, D.A. (1983). *The Reflective Practitioner: How Professional Think in Action*. Aldershot: Arena.

Walklin, L. (1990). *Teaching and Learning in Further Education*. Stanley Thornes (Publishing Limited).

Wilkinson, S. and Hoxley, M. (2005). *The Impact of the 2001 RICS education reforms on the construction industry*. In: A.C. Sidwell (ed.). *The Queensland University of Technology Research Week International Conference 2005, Brisbane*, 4–5 July 2005. Brisbane: QUT.

Additional Reading

Allan, J. (1999). *Actively Seeking Inclusion: Pupils with Special Need in Mainstream Schools*. London: Farmer Press.

Helsby, G. (1999). *Changing Teachers' Work*. Buckingham: OU Press.

Huberman, M. (1995). Professional career and professional development. In: T.R. Guskey and M. Huberman (eds.), *Professional Development in Education* (pp. 21–25). New York: Teachers College Press.

Lovett, H. (1996). *Learning to Listen- Positive Approaches to People with Difficult Behaviour*. London: Kingsley.

Pinar, W.F. (2004). *What is Curriculum Theory?* London: Lawrence Erlbaum Associates.

Tummons, J. (2009). *Curriculum Studies in the Lifelong Learning Sector*. Exeter: Learning Matters.

Wright, A.M., Abdi-James, S., Colquhoun, S., Speare, J., and Partridge, T. (2206). *FE Lecturer's Guide to Diversity and Inclusion*. London: Continuum.

13

A History of Value in the Built Environment

13.1 Introduction

The following chapter charts the contemporary rise of value-driven behaviour. Over its lifetime, the concept has gone by many names. From responsible business, to ethical behaviour, to sustainability, staff well-being, corporate social responsibility, environmental, social and governance, to social value. The concept may have evolved in title, and the focus may have shifted from one pressing concern to another, but the core principles remain consistent in that a business needs to do the right thing. We are now at a stage in the concept's evolution where the idea of Total Value is gaining traction. It is no longer acceptable that an organisation can create or undertake polluting or unethical operations or contribute in any way to any negative issue in society, no matter what 'additional' value-creating behaviour they undertake. An organisation whose very core functions are now viewed as unacceptable and that do not positively contribute to society cannot create and deliver total value. They will be forever undertaking value-driven behaviour in an attempt to 'cover up' the fact that their core business behaviours do not add value to society. The core idea to understand is that the concept of business responsibility is like a bus where nothing ever gets off! As each decade passes, more and more ideas take shape, and more and more expectation is placed upon the shoulders of companies. However, to fully understand where the concept and idea of value is now, and is heading in future, an understanding of its evolution is required.

13.2 1950s

In the 1950s, the idea of responsible business and corporate citizenship began to take root in modern business thought, sparked significantly by the publication of Howard Bowen's seminal book '*Social Responsibilities of the Businessman*' (Bowen, 1953). This period marked the early emergence of corporate social responsibility (CSR) as a legitimate and necessary concern for business leaders in the post-war industrial landscape, but it was some time yet before the term CSR was formally recognised. Social responsibilities of the Businessman is widely credited as the foundation of the modern value-driven movement. Bowen argued that large organisations had a moral obligation to consider the social and economic consequences of their decisions beyond profit generation. He posed a fundamental question: '*What responsibilities to society may businessmen reasonably be expected to assume?*' At a time when business success was measured almost exclusively in terms of financial return, Bowen introduced the groundbreaking (at the time) idea that organisations were not only economic actors but also social institutions with far-reaching influence on the lives of individuals and communities. Largely, this was built on the back of a divided society in the UK and Europe. Post World War II, many countries (and even more economies) had been somewhat decimated by the War, and so governments generally had no money to spend on public services. Yet industrialists, and some organisations, had money. They had supplied a wealth of resources to Governments in fuelling the War efforts, and were rewarded for doing so. What Bowen was calling for in his book was that those individuals who had the means to help society now had the responsibility to help.

Bowen emphasised that organisations could not operate in isolation from society. Their decisions impacted employment, community well-being, education, environmental conditions and overall social stability. As such, they needed to move beyond compliance and actively promote the common good. This idea, although initially met with scepticism by traditional economists and industrialists, began to resonate with a subset of progressive business leaders, particularly in the United States.

Between 1953 and 1959, the socio-political environment further encouraged the idea of responsible business. The post-war boom in the United States brought about unprecedented corporate growth, but also a greater visibility of the power and influence that major organisations wielded. Issues like labour rights, racial inequality and environmental concerns started gaining attention, and critics began holding organisations accountable for their role in contributing to or ignoring these problems. Moreover, during this time, corporate philanthropy slowly began evolving into a more structured form of social responsibility. Organisations started to engage more visibly in community development, support for education and employee welfare initiatives. While many of these actions were still driven by public relations or tax incentives, the broader conversation about the ethical and social role of organisations was clearly gaining momentum.

The term 'corporate citizenship' was not yet commonly used, but the concept was forming. It implied that just like individuals, organisations had duties and responsibilities to society. By the end of the 1950s, business schools and management thinkers had started integrating social responsibility into discussions about corporate ethics and long-term strategy. Though the implementation remained limited and mostly voluntary, the intellectual groundwork had been laid. Bowen's ideas planted the seed for what would later become a global movement, as stakeholders increasingly demanded that organisations be not just economically successful, but also socially and ethically accountable.

However, also during this decade, the seeds of discontent towards the growing social responsibility movement were beginning to be sown. Opposition to such positive social movements began to emerge, with some arguing that social welfare was the responsibility of governments, with the primary aim of organisations being profit maximisation (Levitt, 1958).

13.3 1960s

The 1960s marked a pivotal decade in the evolution of responsible business practices. Building on the foundational ideas introduced in the 1950s, the 1960s saw these concepts expand and gain traction in public discourse, corporate strategies and academic circles. According to Carroll (1999), the 1960s marked a significant period of development in defining CSR, often still referred to then as social responsibility. This era saw several key thinkers attempt to articulate what CSR meant in both theoretical and practical terms.

Davis (1960) defined social responsibility as decisions and actions taken by businessmen for reasons beyond direct economic or technical interests. He acknowledged CSR's complexity but argued it could bring long-term economic benefits. He later expanded CSR to include ethical consequences of actions, stressing that responsibility extends from individuals to institutions and the broader social system (Davis, 1967). Frederick (1960) emphasised that CSR means overseeing the economy in a way that enhances total socio-economic welfare and directs resources toward broad social goals, not just private interests. McGuire (1963) built on this by arguing that organisations have responsibilities beyond legal and economic obligations, including involvement in politics, education, employee welfare and justice. Davis and Blomstrom (1966) went further and argued that CSR was defined as an obligation to consider the broader social system and those affected by business decisions. Walton (1967) added that CSR reflects the close relationship between organisations and society. He stressed voluntarism, corporate ties to other voluntary organisations and the acceptance of costs that may not yield direct financial returns. Together, these thinkers laid the intellectual foundation for modern CSR, expanding the scope of business responsibility and emphasising its ethical, social and systemic dimensions.

During this era, major social and political transformations, including the civil rights movement, the rise of consumer advocacy, anti-war activism and environmental awareness, created new expectations for business. Organisations could no longer operate in isolation from the broader issues affecting society. Instead, they were increasingly held accountable for their impact on workers, communities, consumers and the environment. Companies began to recognise that their legitimacy depended not only on financial performance but also on how they responded to societal needs and demands. The concept of 'social value' emerged more clearly in this context, referring to the wider benefits a company could bring to society beyond its economic output. This included job creation, equitable treatment, ethical governance, community investment and environmental stewardship.

By the mid-1960s, CSR began to evolve from a moral or philosophical concept into a more structured set of practices. Researchers began to formalise CSR as an area of business study, focusing on how companies should balance profitability with ethical conduct and responsiveness to social issues (Carroll, 1999). Large organisations, especially those in industries with significant environmental or social footprints, began incorporating CSR initiatives into their public image and operations. These early efforts were often modest, such as charitable donations, employee welfare programmes or community engagement, but they marked a shift toward the belief that companies had obligations beyond their shareholders.

The environmental movement of the 1960s was a key driver in expanding the definition of responsible business to include ecological concerns. The most influential milestone in this regard was the publication of *Silent Spring* by Rachel Carson in 1962. Carson, a marine biologist and conservationist, meticulously documented the harmful effects of pesticides, especially DDT, on ecosystems, wildlife and human health. Silent Spring was revolutionary not only for its scientific rigour but also for its accessibility to the general public. The book sparked widespread concern about pollution, toxic chemicals and the unchecked use of industrial technology. It directly challenged both government and corporate practices, accusing them of prioritising short-term economic gain over long-term environmental and public health (Carson, 1962). Carson's work is often credited with launching the modern environmental movement. It prompted significant public debate, led to regulatory reforms (including the eventual banning of DDT in many countries) and laid the groundwork for the establishment of the US Environmental Protection Agency in 1970. For organisations, Silent Spring marked a turning point. Environmental responsibility was no longer a fringe concern; it was fast becoming a mainstream expectation. Organisations in agriculture, chemical production and manufacturing faced increasing scrutiny, while others began to see environmental sustainability as a strategic and reputational imperative.

By the late 1960s, CSR was no longer just about doing good; it was becoming integrated into risk management, brand identity and long-term business strategy. Issues such as racial equality, gender equity, consumer protection and environmental conservation became intertwined with corporate reputation and stakeholder trust. At the same time, critics began questioning the sincerity of corporate CSR efforts. Some viewed early CSR as reactive and superficial, as a means to deflect criticism rather than make meaningful change. Nonetheless, the seeds for more systematic and accountable CSR frameworks were sown.

13.4 1970s

The 1970s was a transformative decade in the evolution of good business practices. The growing social awareness and environmental consciousness that began in the 1960s continued to intensify, leading to stronger calls for businesses to act ethically, transparently and sustainably. During this period, the idea of 'doing good' as a business imperative was both challenged and advanced, laying critical foundations for modern responsible business practices. The 1970s opened with a widely debated challenge to the rising CSR movement. The economist Milton Friedman published an influential article in *The New York Times Magazine* titled *The Social Responsibility of Business Is to Increase Its Profits* in which he argued that corporate executives had a fiduciary duty to maximise shareholder returns, not to solve social problems or engage in philanthropy using corporate funds (Friedman, 1970).

Friedman's view reinforced a traditional economic perspective: organisations existed primarily to serve shareholders, and social issues should be addressed by governments or individuals, not organisations. His critique slowed the momentum of CSR in some sectors and sparked an ongoing ideological divide between shareholder primacy and stakeholder responsibility. However, Friedman's perspective also prompted a stronger counterargument in that organisations operate within society and, therefore, have obligations to it. Throughout the 1970s, many companies and thinkers continued to push for a broader vision of organisational responsibility that extended beyond profits.

The environmental movement, energised by Rachel Carson's *Silent Spring* in the previous decade, achieved major legislative victories in the 1970s. The first Earth Day was held in 1970, and the same year, the US government established the Environmental Protection Agency. In the UK, this was followed by key legislation such as the Prevention of Oil Pollution Act (1971), the Nature Conservancy Council Act (1973), which established the Nature Conservancy Council and the Control of Pollution Act (1974).

For organisations, these laws represented a new era of accountability. Environmental compliance was no longer voluntary or symbolic; it was now mandatory, enforceable and increasingly tied to public reputation and legal responsibility. Organisations had to rethink waste disposal, emissions, resource use and environmental impact, often investing in new technologies or practices to meet regulatory standards. While some organisations resisted or attempted to bypass regulations (as many often do), others began to integrate environmental stewardship into their operations, recognising that sustainability could align with long-term business viability. This led to a rise in stakeholder theory, which emerged as a clear challenge to shareholder-focused thinking. Stakeholder theory argues that businesses should consider the interests of all parties affected by their operations, not just shareholders, but also groups such as employees, customers, suppliers, communities and the environment. This shift encouraged companies to assess their social impact more holistically. The idea that businesses should give back to society became more widespread, with a growing number of firms engaging in philanthropy, community development and workplace reforms such as diversity and inclusion initiatives.

The 1970s also saw multinational organisations come under scrutiny for their operations in developing countries, particularly in relation to labour conditions, environmental degradation and political interference. The emerging concept of 'business ethics' began to influence organisational decision-making, and academic institutions started introducing courses in business ethics and CSR to meet the emerging and growing needs of organisations. Public opinion increasingly demanded transparency, fairness and integrity from organisations, especially those operating across borders. This decade saw the early beginnings of organisational codes of conduct and the first attempts at socially responsible investing, as investors sought to support organisations that aligned with their values.

By the end of the 1970s, CSR had evolved beyond charity and compliance. The concept grew to one of utmost importance to all organisations (Wood, 1991). Responsible business was beginning to be seen as a strategic issue tied to risk management, reputation and long-term success. Although there was no universally accepted model of CSR at the time, organisations increasingly understood that ethical lapses, environmental negligence or social disregard could damage their brand and threaten profitability.

13.5 1980s

The 1980s marked a dynamic and often contradictory decade in the evolution of good business practices and CSR. There was Margaret Thatcher in the UK and Ronald Regan in the USA, with both leaders arguably well known for economic liberalism and deregulation. A global economic decline also led organisations to refocus on profit as a core business strategy (Carroll, 1999). However, there was also rising global awareness of social and environmental issues, alongside new pressures from consumers, investors and activists. Organisations were therefore pushed to take greater responsibility for their actions, both economically and ethically. The tension between profit-driven strategies and ethical obligations defined the CSR discourse of the 1980s.

The decade began with a strong emphasis on free-market capitalism. Economic policies promoted deregulation, privatisation and reduced government intervention. The idea that the primary purpose of business was to generate profit for shareholders, which originated in the previous decade (Friedman, 1970), gained significant traction. This era saw aggressive corporate takeovers, layoffs and cost-cutting strategies aimed at maximising shareholder value. Within this context, CSR often took a back seat to financial performance. Many organisations often focus on short-term gains rather than long-term social investment. However, this environment also gave rise to critical questioning of corporate ethics and social obligations. As scandals and controversies emerged and public scrutiny of organisational behaviour intensified.

The 1980s witnessed several major environmental disasters that highlighted the risks of corporate negligence and revealed the legislation and regulations that had been introduced to be weak and often with little enforcement capability. In response to the disasters and growing public concern, there was an emergence of ethical consumerism. Consumers began demanding more information about how products were made, where they came from and whether organisations upheld ethical labour and environmental practices. Activists and NGOs increasingly held organisations accountable through public campaigns, shareholder resolutions and boycotts. This activism pressured many businesses to adopt codes of conduct and improve transparency. Issues like sweatshop labour, pollution and corporate governance became central to the public's assessment of good business.

Despite the dominant narrative of deregulation and profit maximisation, the 1980s also witnessed the institutionalisation of business ethics and the growth of strategic CSR. Popularised by the Brundtland Report (1987), which brought worldwide attention to the idea of sustainability and encouraged a shared understanding to emerge of a previously ambiguous concept (it defined sustainability as meeting the needs of the present without compromising the ability of future generations to meet their own needs). Rather than viewing CSR as philanthropy, some companies began to align social initiatives with business objectives. This meant investing in employee well-being, supporting local communities and reducing waste. For altruistic reasons as well as because such actions enhanced brand reputation, customer loyalty and employee engagement. The decade's lessons set the stage for the more formal and globally integrated CSR frameworks that would emerge in the 1990s and beyond.

13.6 1990s

The 1990s marked a critical turning point in the evolution of good business practices, as CSR, sustainability and ethical governance moved from the margins into mainstream business thinking. Globalisation, the rise of digital communication and increasing public scrutiny made it clear that organisations could no longer operate in isolation from social, environmental and ethical issues. The decade saw the formalisation of CSR strategies, the birth of global sustainability frameworks and a growing awareness that long-term success required alignment with stakeholder interests and societal values. The decade was defined by rapid globalisation and the mainstream reporting of CSR as a part of everyday business identity (KPMG, 2017). Organisations expanded across borders, supply chains became more complex and developing markets became increasingly important. While globalisation brought economic opportunity, it also highlighted deep social and environmental inequalities. High-profile controversies involving labour exploitation, deforestation and human rights abuses, often involving large multinationals, led to growing public demand for responsible business practices. Consumers, NGOs and the media began holding organisations accountable not just for what they sold, but how they produced it and where it was produced. Global operations meant global responsibilities, and organisations started to realise that poor practices abroad could have serious reputational consequences at home.

At this time, CSR matured from a set of ad hoc charitable activities into a more strategic, integrated part of everyday operations. Many large organisations established formal CSR teams and began publishing CSR or sustainability reports to document their social and environmental performance. Rather than viewing CSR as a cost, organisations increasingly saw it as an opportunity to build trust, reduce risk and differentiate themselves in competitive markets. The notion of the 'triple bottom line (people, planet and profit)' gained popularity, emphasising that business success should be measured not only by financial performance but also by social and environmental impact.

Key events such as the 1992 UN Earth Summit in Rio de Janeiro were a landmark event that brought together governments, NGOs and organisations to discuss sustainable development. The summit introduced Agenda 21, a global action plan for sustainability, and underscored the importance of integrating environmental and social concerns into decision-making. Agenda 21 was ultimately a precursor to and replaced by the Sustainable Development Goals. The Global Reporting Initiative was launched in 1997 and began setting standards for CSR reporting, helping companies better measure and communicate their non-financial performance. Movements like that of 'Fair Trade' also grew significantly, as consumers demanded ethically sourced products, particularly in food and textiles. This reflected a broader rise in ethical consumerism, where buying decisions were influenced by social and environmental values. The 1990s also saw a proliferation of codes of conduct, particularly in industries like apparel, technology and mining. These codes set minimum standards for labour practices, environmental protection and ethical sourcing. At the same time, corporate governance became a focus of concern following several financial scandals and governance failures. Organisations were expected to not only obey the law but also act with transparency, fairness and accountability to all stakeholders, including shareholders, employees, communities and the environment.

Finally, the rise of the internet and 24/7 global media in the 1990s meant that information, either good or bad, spread quickly. This new transparency empowered activists and watchdog organisations to expose unethical practices and amplify consumer voices. Organisations learned that their reputations could be damaged overnight, creating strong incentives to adopt proactive and authentic CSR strategies. Good organisational practices, therefore, were essential not just for reputation, but for long-term resilience and success. This set the stage for the sustainability revolution of the 21st century.

13.7 2000s

Between 2000 and 2009, the evolution of good business practices accelerated dramatically, as CSR (the term was still widely used at this point), sustainability and ethical governance became central to business strategy rather than optional add-ons. This decade witnessed a growing integration of social and environmental concerns into core business operations, driven by stakeholder pressure, regulatory changes, financial incentives and the increasing visibility of corporate behaviour in a globalised and digitally connected world. It was a period of both innovation and accountability that laid the groundwork for modern ESG (Environmental, Social and Governance) frameworks.

By the early 2000s, CSR had matured into a recognised component of global business strategy. Organisations realised that responsible practices not only enhanced reputation but also attracted investors, customers and top talent. The concept of the Triple Bottom Line (people, planet, profit), which gained traction in the 1990s, became a guiding framework. However, it was increasingly refined into measurable, reportable categories that would later evolve into what we now call ESG metrics. The idea that long-term value creation required attention to non-financial factors, like climate change risk, labour practices, community relations and ethical governance, became increasingly accepted by forward-thinking companies and investors alike. One of the key developments in this era was the rise of standardised reporting frameworks. The GRI became widely adopted, helping organisations report on sustainability issues with greater transparency and consistency. In 2000, the United Nations Global Compact was launched. This voluntary initiative encouraged businesses to align their operations with 10 universal principles covering human rights, labour, environment and anti-corruption. It marked a significant step in recognising that business is a vital partner in achieving global development goals. In 2006, the UN also introduced the Principles for Responsible Investment (PRI), which asked institutional investors to incorporate ESG issues into their investment decisions. This initiative marked the start of mainstream integration of sustainability into global capital markets.

The 2000s also brought numerous corporate scandals that reshaped the way organisations approached governance and ethics. High-profile failures both before and during the global financial crisis highlighted the risks of poor governance, lack of transparency and excessive short-termism. These events led to stricter regulations which enforced accountability in financial reporting and corporate governance. The fallout from these crises reinforced the idea that ethical behaviour and good governance were essential to sustainable business performance. The early 2000s also saw climate change become a defining issue for organisations. Growing scientific consensus and high-profile events pushed climate risk into the public and corporate consciousness. In response, many organisations began measuring their carbon footprint, setting emissions reduction targets and investing in renewable energy. Sustainability was no longer just a reputational tool; it became a strategic priority. Initiatives like the Carbon Disclosure Project and the rise of green certifications for buildings helped organisations benchmark and reduce their environmental impact.

During this decade, consumer expectations shifted dramatically. The rise of the ethical consumer, someone who values fair trade, sustainability and ethical supply chains, pressured organisations to become more transparent about their sourcing and labour practices. This trend also saw the growth of social enterprises and B Corporations, businesses that embedded social and environmental goals into their legal and operational structures. These models blurred the line between profit and purpose, offering alternatives to traditional corporate structures. The 2000s were also defined by the explosion of digital technology and social media, which transformed how organisations interacted with the public. It also transformed the way members of the public could keep track of organisations, and how they could become aware of any indiscretions happening in other locations or further existing supply chains. Information became instantly shareable, and reputational risks increased exponentially. Corporate missteps, whether environmental, ethical or social, could (and did) go viral and damage brands within hours. This digital transformation also empowered all stakeholders to demand greater accountability, pushing organisations to maintain high standards and transparency. The rise of ESG frameworks, growing climate awareness, high-profile corporate failures and empowered consumers pushed companies toward greater responsibility and transparency. This decade laid the groundwork for the deep embedding of sustainability into business models, making ethical and responsible behaviour not just desirable – but essential – for long-term success.

13.8 2010s

From 2010 and 2019, the evolution of good business practices underwent a significant transformation, becoming deeply integrated into corporate strategy, risk management and stakeholder engagement. This decade marked the mainstreaming of sustainability, CSR and ESG practices. Although the term CSR was becoming quite dated by this point and would largely fall out of common and contemporary organisational lexicon by the end of the decade. Organisations moved beyond simply managing reputation and compliance toward creating long-term value for society, the environment and shareholders. With growing global challenges like climate change, inequality and political instability, organisations increasingly recognised their role in solving systemic problems alongside governments (and sometimes instead of).

One of the most defining trends of the 2010s was the rapid rise of ESG investing. Investors, particularly large institutional ones, began incorporating ESG criteria into financial decision-making. It was now common practice for all large and medium-sized organisations, both public sector and private sector, to produce annual ESG reports. The UN Principles for Responsible Investment (PRI) gained widespread adoption during this decade, and ESG performance started to influence stock prices, shareholder voting and access to capital. Sustainability was no longer just a moral imperative; it was fast becoming a financial one. The introduction of the Public Services (Social Value) Act (2012) served to formalise (and regulate) organisational action when it came to ESG behaviour. The Act mandated that all public sector procurement over a set threshold must include non-financial 'social' based procurement criteria, and that public sector contractors could no longer be awarded on a purely lowest cost basis. It was becoming formally recognised that a focus on short-term cost was detrimental to the achievement of long-term value. The introduction of the Social Value Act and the general movement of ESG (and CSR before it) have arguably served to refocus the idea of Friedman's (1970) that an organisation's focus should be on shareholder profits. If this idea is true, then it is now imperative that organisations consider value-based behaviour, as it is only by engaging with value-based behaviour that an organisation can be successful in procurement and thus make profitable returns for shareholders.

In 2015, the United Nations launched the Sustainable Development Goals. These are a set of 17 global goals to end poverty, fight inequality and tackle climate change by 2030. The SDGs provided a common language and framework for businesses to align their strategies with global development objectives. Many organisations began mapping their initiatives on the SDGs, integrating them into sustainability reports and using them to communicate purpose and impact. The SDGs reinforced the idea that good business practices must deliver value not only to shareholders but also to society and the planet. It also made the idea of value behaviour global and served to install benchmarks by which organisations and governments could be measured against. The Paris Agreement, adopted in 2015 by 196 countries, aimed to limit global warming to well below 2 °C, with efforts to cap it at 1.5 °C. This international accord significantly shaped business practices, particularly in carbon-intensive industries, and installed climate change as one of the defining environmental challenges of the decade, bringing it to the forefront of many organisations' value-based behaviours. Organisations began setting science-based targets to reduce greenhouse gas emissions, investing in renewable energy and committing to net-zero goals. Climate risk disclosures also became more sophisticated, with frameworks like the Task Force on Climate-related Financial Disclosures encouraging organisations to assess and report how climate risks affect their operations and financial health.

Alongside environmental concerns, the 2010s saw heightened attention to human rights, labour practices and equity, diversity and inclusion. The #MeToo movement, which gained momentum in 2017, brought workplace harassment and gender inequality into sharp focus, forcing companies to review policies, culture and leadership structures. Supply chain ethics also came under scrutiny, with growing expectations for businesses to ensure that products were not linked to forced labour, unsafe working conditions or human rights abuses. As a result, more organisations adopted modern slavery statements, strengthened supply chain audits and implemented responsible sourcing policies.

Digital transformation and the evolution of social media continued into this decade and further empowered stakeholders to demand transparency and accountability. Information about an organisation's actions spreads globally in real time (and often quicker than an organisation could respond to). This constant visibility meant businesses needed to be more proactive, authentic and evidence-driven in their ESG efforts. Technology also enabled more accurate data collection and reporting on sustainability metrics. Companies could now monitor emissions, track supply chains and evaluate social impact with greater precision, helping build trust with consumers, investors and regulators. Equally, organisations had no reason not to, with the technology to track and communicate the impact of value-based behaviours readily available to all organisations. This decade laid the foundation for a business environment where doing good and doing well became inseparable.

13.9 2020s

From 2020 to the present day (2026), the evolution of good business practices has accelerated and deepened in response to unprecedented global challenges. This period has been marked by a convergence of crises (such as the COVID-19 pandemic, rising climate emergencies, social justice movements, economic instability and geopolitical tensions) that have profoundly reshaped how businesses define responsibility, resilience and value creation. As a result, concepts such as sustainability, ESG, corporate purpose and stakeholder capitalism have evolved from aspirational ideas into operational imperatives.

The COVID-19 pandemic, beginning in early 2020, served as a global wake-up call for organisations. It exposed vulnerabilities in global supply chains, labour practices and healthcare systems and forced organisations to re-evaluate their social responsibilities. All organisations were now judged not just by their financial resilience, but by how they treated employees, supported communities and responded to public health needs. This period saw the emergence of 'responsible resilience' as a core business theme, highlighting the importance of human capital, mental health, flexible work and community engagement. Remote work, digital transformation and inclusive leadership became central to business continuity and innovation.

So far this decade, ESG investing has surged. Asset managers, investors and regulators place growing emphasis on non-financial performance. ESG ratings and disclosures are becoming critical to investor decision-making, and organisations are facing rising pressure to demonstrate measurable action on climate, equity and governance issues. The climate crisis remains a dominant focus. As extreme weather events increase in frequency and severity, more companies are committing to net-zero targets, science-based emissions reductions and climate risk reporting. Indeed, to not have such targets these days, an organisation would be seen as an outlier amongst its peers.

In the post-pandemic world, organisations increasingly recognised that ESG risks are interconnected. Climate change, biodiversity loss, inequality and political instability are not isolated challenges but systemic threats to business continuity. This shifted the role of business from being part of the problem to being a key part of the solution. Forward-looking organisations embed sustainability into corporate strategy, innovation pipelines, risk management and finance. Organisations now see value-based behaviour as their primary purpose. Where this is not the case, value-based behaviour is integrated and embedded across every aspect of business operations.

13.10 Conclusion

Without having a clear and tangible focus on value creation and delivery, questions will be asked in society as to what the true benefit of an organisation is, and whether alternatives exist that can fulfil the same primary business need and purpose, but that result in simultaneous value creation when used. It is no longer acceptable that organisations simply exist, even if they are providing a service to society. All organisations now also need to create value. The idea of value, which first emerged under the guise of 'business responsibility' in the 1950s, has been on an evolutionary journey since this time. However, like the metaphorical 'bus' introduced at the start of this chapter, the concept of business responsibility has simply increased in expectation as the decades have passed. Each new decade brings new ideas and expectations. None of which have 'left the bus', and all of which now sit together in the demands they place upon organisations and the expectations society has upon business. It is no longer enough to continue business operations as normal and treat value-based behaviour as an additional organisational operation. Value needs to be intertwined with all operations. A failure to do so will lead to society viewing an organisation as outdated and obsolete, with no place in modern society.

References

Bowen, H.R. (1953). *Social Responsibility of the Businessman*. New York: Harper & Row.

Brundtland, G.H. (1987). *Our Common Future. Report of the World Commission on Environment and Development*. United Nations.

Carroll, A.B. (1999). Corporate social responsibility: evolution of a definitional construct. *Business & Society*, 38: 268–295.

Carson, R. (1962). *Silent Spring*. Houghton Mifflin.

Davis, K. (1960). Can business afford to ignore social responsibilities? *California Management Review*, 2(3): 70–76. https://doi.org/10.2307/41166246.

Davis, K. (1967). Understanding the social responsibility puzzle. *Business Horizons*, 10(4): 45–50.

Davis, K. and Blomstrom, R.L. (1966). *Business and Its Environment*. New York: McGraw-Hill.

Frederick, W.C. (1960). The growing concern over business responsibility. *California Management Review*, 2(4): 54–61.

Friedman, M. (1970). The social responsibility of business is to increase its profits. *The New York Times Magazine* (September 13), 32–33, 122–124.

KPMG (2017). *The Road Ahead: The KPMG Survey of Corporate Responsibility Reporting 2017*. KPMG Global Sustainability Services.

Levitt, T. (1958). The dangers of social responsibility. *Harvard Business Review*, 36: 41–50.

McGuire, J.W. (1963). *Business and Society*. McGraw-Hill.

Walton, C. (1967). *Corporate Social Responsibilities*. Belmont, CA: Wadsworth.

Wood, D. (1991). Social issues in management: theory and research in corporate social performance. *Journal of Management*, 17(2): 383–406.

14

Practitioner and Student Examples

14.1 Introduction

In this chapter, specific scenarios and questions are posed to both Built Environment industry professionals and students (of all levels) with regard to achieving total value. Whilst these scenarios are all fictitious examples, they reflect real-life situations and illustrate the broad range of value-based decision-making that is required to transform the industries that operate within both the Natural and the Built Environment into value leaders.

This chapter reveals that value-driven procurement is often the catalyst for value delivery. Whilst value delivery can still be achieved even where there are procurement requirements, it is apparent that a top-down approach to encouraging and supporting value-based behaviours is the most effective and efficient way to achieve a high level of value delivery. Sometimes a specific form of value may be delivered, in other examples, the word 'value(s)' is utilised as the value that can be delivered is broad and can encompass many different forms.

14.2 Scenarios

Scenario 1

The National Health Service (NHS) Trust wants to procure a range of management consultancy services in the first step towards an ongoing broader procurement exercise that will be looking to invest in several local healthcare building redevelopments over the next decade. The Trust has little knowledge of construction or value creation and wants the management consultancy to help advise on procurement and on maximising value at every step of the redevelopment process.

Question 1.1
Can the NHS Trust use value-based criteria in its procurement of the consultancy services despite having little knowledge of value creation at this stage?

Answer 1.1
The short answer is yes, even to procure the management consultants who will be advising on how best to maximise value, the NHS client can still consider how best value can be created at this early stage. If, in this instance, the NHS Trust does not have the knowledge and experience of what to ask for with regard to value creation, they can ask the tendering management consultants for their own suggestions. For example, the tender criteria used to procure the management consultants could ask what social value each management consultant proposes to be achieved, both in the future redevelopment of buildings and in the immediate procurement of their organisation. Alternatively (or additionally), the NHS Trust could also ask how each of the procuring management consultants will align their value creation with a recognised and established value framework (as discussed in Chapter 5).

Question 1.2
Is the NHS Trust bound by any legal or regulatory statutes to consider value creation in the building redevelopments?

Answer 1.2
Yes. The Public Services (Social Value) (2012) places a legal duty on all public sector bodies (such as NHS Trusts) to consider how social value can be maximised during procurement, with the Act stating that such value considerations should be made during procurement, and the awarding of any projects should not be on a purely cost-based approach. The Act is relevant for any services procured over a set threshold but is often adopted as standard practice across all public body tenders. There is also the Procurement Act (2023) and many different Public Procurement Notices that all dictate the practices that must be adopted, elements that must feature and mandatory requests to be made. In short, as the NHS is a major public sector client and often delivers complex mega projects. It is under constant scrutiny from a diverse range of stakeholders and therefore needs to clearly evidence of compliance with an extensive range of demands.

Question 1.3
What are the best value-based solutions that meet the needs of the Trust?

Answer 1.3
This is often a two-pronged approach. Firstly, a client will know what sort of value they want to achieve, or the stakeholders they wish to feel a positive impact on the value creation. Even if the client is unable to clearly articulate the exact expected outcomes or the best methods of achieving any desired value. The second prong is where an effective management consultant will utilise their knowledge and research ability of the local area to help inform the client of other areas of potential positive value creation that the client may be unaware of. Therefore, the best value-based solutions to meet the NHS Trust's needs in this scenario will be a combination of the existing locally based needs the Trust will have some knowledge of, and the research-informed needs the management consultant will be aware of.

Scenario 2

A main contractor is tendering for a £10 m private sector mixed-use residential and commercial development. The client is a speculative private sector developer, and the proposed site location is in the South of England.

Question 2.1
Is the client in this example bound by the Public Services (Social Value) Act (2012)?

Answer 2.1
No. As a private sector client, they are not bound by the Public Services (Social Value) Act (2012). However, they can extend their procurement criteria beyond the traditional focus on time, cost and quality and include all manner of value-based criteria if they wish to generate more value as a result of their procurement spending. However, legally, the private sector client must still ensure all their practices are clear, transparent and open for scrutiny with adequate rights to appeal. This is not to say they have to share all details of the procurement, and the client is still able to maintain a high degree of commercial sensitivity, in order to prevent the loss of any commercially sensitive materials. Nevertheless, the private sector clients must have a fair procurement process and can decide to include value-based procurement criteria if they choose to do so.

Question 2.2
Can the main contractor still provide value-based services over and above the procurement requirements stipulated by the client?

Answer 2.2
Yes, the main contractor can still provide any elements of value creation they wish. However, unfortunately, in this situation, such value is unlikely to be recognised by the client. Therefore, the main contractor will still have to base their tender on the time, cost and quality focus of the client in order to try and achieve procurement success. In such situations, all value created should still be measured by the contractor. Firstly, this enables them to communicate such value in both project-specific case studies and also in any annual reports they produce. Secondly, evidence of such value can then become of paramount importance in helping to evidence their value creation experience for any future tender submissions, particularly for clients with a stronger focus on value creation.

Question 2.3
Will the geographical region have any influence on the type of value that can be created if either the client or make contractor wanted to consider value creation?

Answer 2.3
Yes. Value creation very much has a local focus and is specific to whichever region the development is undertaken. Whilst national trends, needs and best practice examples can be relevant and serve as an important source of lessons learnt, any value-focused behaviour should be anchored in the local needs of the relevant geographical region.

Scenario 3

An experienced private sector developer based in Scotland wants to adopt a value-based approach to a small private sector housing development during the procurement of a main contractor. However, the developer is unaware of how value can be created, and also unsure if their value-based ideas can be enforced with the main contractor, as everything they undertake is private sector work.

Question 3.1
Can the private sector developer enforce value creation despite not being a public sector body?

Answer 3.1
Yes. Whilst the private sector developer is under no legal obligation to include value-based criteria or requirements, they can definitely include value-based criteria in their tender documents. If they clearly illustrate the importance in the tender documents, then all those contractors who are completing the tender returns will know to include value delivery. The types of value delivered could be broad, and it depends on how the tender documents are drafted as to what sort of value-based behaviour is encouraged. Generally, private sector clients are not as focused on value-driven behaviour as public sector clients. Although a few socially and environmentally driven and conscious private sector clients are challenging this idea and changing the 'mould' for what private sector construction procurement looks like, and the wider societal benefits that it can achieve.

Question 3.2
Can the private sector developer dictate a focus on value created only in Scotland?

Answer 3.2
This can be achieved, but careful wording will need to be used in all procurement documentation. Again, as a private sector organisation, there is no obligation upon them to include value creation as part of the tender requirements. However, if this is important to the client, value-based procurement criteria can be included. It is the 'hyper local' approach that may cause issues. It is increasingly difficult to specify criteria requirements that may be exclusionary in their nature to some groups over others, almost as a form of positive discrimination. Some clients may ask (for example) for supply chain spend percentages to be within a set mileage radius of a selected location. Others may ask for certain demographic groups to be targeted for employment and training initiatives. Such targeted initiatives can be requested, but clients must be cautioned about going 'too far' with any groups they inadvertently exclude, as this could lead to challenges and issues further down the construction lifecycle.

Question 3.3
Why would the private sector developer want to focus on value creation, perhaps at the expense of profit maximisation, when they are not legally obliged to?

Answer 3.3

There are many reasons why this might be the case, but all will ultimately come down to the owner/higher management's personal values. There is ample research that evidences how the values of ownership will inevitably influence and direct the values of the organisation. Essentially, in this instance, the organisation (and by extension the ownership) is more driven by their values than the need for profit. This is not uncommon, and many organisations across the UK and internationally have historically been driven by more altruistic motives. Often (and especially in contemporary society), value creation closely aligns with profit generation, with socially and environmentally aware consumers more likely to pay a premium for a product or service that delivers broad value. Even where organisations still retain a focus on profit maximisation, understandings are increasing that in order to continue to maximise profit, value-based behaviours need to be adopted, and value needs to be created in order to win work. However, there still remain many organisations that are committed to creating value, even at the expense of making additional profit. An organisation may pay sick pay or maternity pay far in excess of the minimum legally required. They may provide additional holiday days or other benefits that essentially cost the organisation from its own bottom line but are benefits the organisation continues to offer through a sense of value-driven ethics that are intertwined with the owner/leadership/organisation's standard way of operating.

Scenario 4

A public sector client is procuring for a housing development partner and wishes to utilise a Modern Methods of Construction manner of delivery, with a focus on prefabricated housing solutions.

Question 4.1

Can the public sector client adopt a value-based approach whilst also adopting a Modern Methods of Construction approach?

Answer 4.1

Yes. But this is increasingly difficult in the modern climate. There is somewhat of a disconnect between two competing agendas in (1) modern methods of construction and (2) broader values such as social value. At present, the public sector is more focused on achieving social value (largely due to its legislative requirements), whereas the practices and processes involved in modern methods of construction are optional. However, this 'option' is rarely selected due to its innovative and therefore inexperienced delivery. This is especially pertinent in a public sector procurement scenario where public sector money is being spent, and therefore, value for money needs to be evidenced. There is something increasingly difficult to do with innovative and largely untested methods. Notwithstanding the lack of trust and certainty afforded to innovative delivery methods such as modern methods of construction, these methods can be combined with value-based decision-making. Firstly, the adoption of modern methods of construction brings with it the achievement of broad values. For example, sustainable value can be achieved as there is a reduced use of resources and raw materials and less waste. Secondly, values such as social value can still be achieved, but it would require a new perspective from the public sector body. At present, social value has a 'hyper-local' focus, but the use of modern methods of construction processes serves to remove this hyper-local focus as the value creation is concentrated in the manufacturing region and not necessarily in the construction region. A re-consideration of what social value means could align with the decision to utilise modern methods of construction and achieve broad value outcomes.

Question 4.2

Would adopting a Modern Methods of Construction approach create more or less value than adopting a traditional construction approach?

Answer 4.2

There is no clear answer to this question as the circumstances will differ significantly from one project to the next. However, it is essential to understand that a clear and informed comparison needs to be undertaken to arrive at the correct answer for each scenario. In this context, the client is a public sector body, and so if a local body, a focus on Modern Methods of Construction processes may reduce the local value otherwise experienced, such as the local social value generated from employment and the local economic value generated from local spend. However, such an approach may increase local environmental value from less project site traffic and machinery use, as many building components will be assembled off-site before being transported and installed in a shorter duration of site works. A national public sector client could therefore take a more geographically spread view that includes the increased social and economic value in one area and the enhanced sustainable and environmental value in another and claim them all as success under one project. However, a local public sector client will often be limited in their geographical region and therefore be accountable for all value types maximised within that region. This could lead them to make decisions that maximise the value types they are most familiar with, and also potentially make decisions that maximise the value their region experiences, even at the expense of value creation elsewhere, which may ultimately lead to lower levels of national and international value being created as a whole. With the adoption of Modern Methods of Construction processes specifically, the opportunity to create more value exists, although if this opportunity is more valuable nationally, the processes may not be adopted, as a local agenda may triumph over a national one.

Scenario 5

An internationally based private sector main contractor is looking to submit a series of tenders for both public sector and private sector construction work in the UK. They specialise in several markets, including student accommodation, residential and commercial, but have not previously attempted to win any UK-based work.

Question 5.1

How would the internationally based main contractor best evidence that they can achieve UK-based value?

Answer 5.1

Whilst it may be easy for the contractor to evidence value creation in international markets, generally in UK-based public sector tenders, the procuring body will require information on the value that is to be created within the geographical region applicable to the tender. This could perhaps be UK-wide or could be much more regionally specific, depending upon the project requirements. Procurement documents may be very specific in the type of value they require to be created and delivered. In this case, the international-based contractor knows exactly what value they should be looking to deliver. Whilst their tender return can be supplemented with evidence of previous performance based in other locations, it should be primarily focused on exactly what value they will deliver, and this should match the value specified by the client. Where this is not the case, and value requirements are more flexible and less prescriptive, the international contractor has more free rein but should always remember to follow the same format and avoid focusing too heavily on their international value-based delivery experience (no matter how vast). The same rules should be followed for private sector clients, and the contractor should either be led by the client in the procurement documents as to the value(s) to be achieved, or they should look to lead the client with what can be achieved – but always with a local and forward-thinking focus.

Question 5.2

Would it be acceptable for the main international contractor to create value in one country when delivering services in another country?

Answer 5.2

This would be of benefit globally and align with many aspects and requirements of value frameworks such as the Sustainable Development Goals and their global reach and remit. However, to be successful in the procurement process, a local focus is generally favoured by the procuring organisation. This is often of paramount importance and underpinned by legislative requirements (if, for example, the Social Value Act (2012) has been interpreted in a hyper-local way). Therefore, a local focus should be adopted for procurement purposes. However, at a group organisation level, and as an identity/strategy/purpose-driven approach, international-based value creation should definitely be encouraged. With an increasing global focus on all industries and global problems rising or stagnating across many regions, organisations opting to tackle issues that are close to their base of operations, closely align with their operational practices, or just those causes that are close to the 'hearts' of organisations and their staff, should definitely be encouraged. Such practices would be considered 'best' practices and should be championed to illustrate end evidence what can be achieved by international-based organisations in the hope that others follow suit and begin got tackle those global problems that procurement practices rarely tackle.

Scenario 6

A UK-based public body client is procuring a £50 m office building and has received an intention to submit notices from several internationally based contractors expressing an interest in submitting bids for the project.

Question 6.1

Can the UK-based public body client still achieve maximum value on the project if they are to award the works to an internationally based main contractor?

Answer 6.1

Yes, the maximum value can still be achieved in such a situation. Here, the emphasis and onus should be placed upon the client to be prescriptive in exactly what type of value(s) they are looking to create and buy when they are hoping to be delivered. During the delivery, the contractor can be originally based in any country or jurisdiction, as the work undertaken should be of the same quality and in the same location. Maximum value can still be achieved, as any successful contractor can deliver the same initiatives and create the same value (in theory). It should all come down to which contractor fulfils the majority of the required value-based criteria. Often this may be the local contractors who have local knowledge, but equally international-based contractors may have the skills, resources and experiences to deliver exactly what the client needs. Essentially, the main determining factor in whether or not desired value outcomes and impacts are achieved is the robust measures and governance put in place by the procuring organisation.

Question 6.2

Is the UK-based public body client able to discount the internationally based main contractor from submitting their tender as part of the procurement process if they wanted only UK-based main contractors to submit tenders?

Answer 6.2

This largely depends on the requirements and regulations governing the procuring body. For large contracts that are awarded under European Union (EU) legislation, any organisation that meets the requirements from within the EU is able to register its interest and submit a tender. The UK is no longer bound to abide by this legislation, and so any procurement activity within the UK alone has to abide by UK legislation only, and there is no current requirement for an obligation to include internationally based organisations. However, UK-based public body clients may want to allow any organisation with the required experience and expertise to bid for works to maximise project success. Nevertheless, public sector bodies can set their own procurement requirements (within reason and as long as they abide by all regulations and legislation) and so can dictate the organisations they wish to submit tenders (as long as the public sector body can be open and transparent and justify all choices and rules made against scrutiny and challenge).

Scenario 7

A UK public sector client is running a framework procurement exercise for a consultant partner to provide professional support services. They are inexperienced in framework management and undertaking large-scale procurement exercises.

Question 7.1

The public sector client is requesting value achievement in the procurement but has not specified the type of value they want to be delivered. What will be the potential ramifications of this during the procurement?

Answer 7.1
Without clear procurement guidance on the types of value required, it will be increasingly difficult to review and compare the received tenders. If one consultant returns their tender stating they will deliver one type of value through one activity and another consultant states they will deliver a different value-creating activity, the public sector client will have to compare and rate each tender return. Without clear guidance, they have set themselves over what value they want to achieve and how, it will be very difficult to arrive at a clear answer. Without a clear answer, any end result may be subject to scrutiny and possible legal objections.

Question 7.2
How can the clients' limited experience serve as both a benefit and a drawback during the procurement process?

Answer 7.2
Whilst having clear and extensive knowledge can almost always be beneficial to ensuring maximum value can be achieved, the downside is that it can lead to a lack of innovation and evolution of value creation and delivery. Organisations that lead in value delivery can sometimes succumb to sticking with what their current practices (as they work well) and fall into the trap of not pushing these practices further. Therefore, organisations that have a lack of experience in value creation and delivery may start by asking the basic questions, but then progress from this to challenging currently held fundamental beliefs and ultimately pushing the concept of value creation further on its evolutionary journey. For all parties and stakeholders, it is essential to understand that value expectations, standards, practices and behaviours are always evolving, and it is essential to understand and accept this, and always aim to be at the forefront of the concept to ensure value can be maximised.

Scenario 8

A UK public sector project may have funding from international partners, national funding pots, local grants and the private sector if any elements are delivered as a joint venture.

Question 8.1
Can the public sector body leading this project simultaneously achieve and satisfy all value requirements of all the funding bodies involved?

Answer 8.1
It is possible to simultaneously satisfy all funders' individual requirements, but often these can be competing, and so it may be difficult to fully satisfy all requests and requirements. However, many of these requirements will overlap, and with accurate and robust measurement practices in place, it is possible to use the same initiatives and data for all funders. Caution has to be exercised here to avoid 'double counting', which is where the outcome, outputs and impacts of an initiative are counted more than once to claim greater value has been delivered. Yet it is still possible (and entirely accurate) to share the value impact data with more than one body. Where value requirements do not align (such as the international partner may want to see international-based value creation and cooperation, and local funding partners want to see a high percentage of local spend), the public body will

have to decide how to best allocate its resources amongst the initiatives it could undertake. Any negative implications could be minimised if the public body leading the project has a true understanding and appreciation of the value requirements at the start. These can then be planned for, and clear and consistent communication can occur with all partners regarding value creation expectations.

Question 8.2

Can any joint venture private sector partners be held to account for any of the required value-based deliverables?

Answer 8.2

Yes, if they are procured via a suitable and robust method. A comprehensive framework agreement could be utilised to outline all short, medium and long-term value-based requirements. Having an accurate understanding of these early on will aid the public body leading the project in ensuring the value targets are delivered. Procuring a joint venture partners too early in the process (before value requirements have been truly understood) or procuring the partner with no clear communications regarding value delivery will inevitably lead to a lack of value creation. Often, the problems with achieving the desired level of value, or even with suitably recording and communicating the value delivered, only arise in the latter stages of project delivery, leading stakeholders to think the problems are then due to those involved at the project delivery stage. However, the majority of such problems almost always stem from much earlier in the construction process, often as early as the feasibility stages of a project. Value-based decisions should be made as early in the construction process as possible to allow their requirements and ramifications to be truly understood. In this instance, waiting until the private sector joint venture partner is being procured to consider how value delivery can occur will be too late to suitably ensure all value achievement can be maximised.

14.3 Conclusion

This chapter outlined several different scenarios, all of which are based on true-life situations. The questions posed are ones which must be asked to truly understand and appreciate value creation and delivery, as well as to ensure value maximisation. The answers attempt to answer both the direct question(s) asked and also provide a background rationale as to why there are often no direct right or wrong answers, and in many cases, context-specific information is needed to ensure the best outcome is arrived at. Ultimately, by reviewing the scenarios, considering the questions posed, and then reviewing the answers provided, a better appreciation for value-driven behaviour will be gained. The answers and responses generally may not be agreed with, or some may feel have not gone far enough, but by having a starting point for discussion, the concept of value and the different types of values that can be delivered will all evolve as concepts, and their positive impact will be maximised. Initiatives that are not as effective can then be reflected upon, analysed and, where needed, improved upon to ensure constant and effective value creation and delivery.

Index

m

n

o

www.ingramcontent.com/pod-product-compliance
Lightning Source LLC
LaVergne TN
LVHW080409150526
838656LV00003B/77